Practical Design Calculations

for Groundwater and Soil Remediation

PRACTICAL DESIGN CALCULATIONS

for Groundwater and Soil Remediation

JEFF KUO, PH.D., P.E.
Civil and Environmental Engineering
California State University, Fullerton

LEWIS PUBLISHERS
Boca Raton London New York Washington, D.C.

Library of Congress Cataloging-in-Publication Data

Kuo, Jeff.
Practical design calculations for groundwater and soil remediation / Jeff Kuo.
p. cm.
Includes bibliographical references and index.
ISBN 1-56670-238-0 (alk. paper)
1. Soil remediation—Mathematics—Problems, exercises, etc.
2. Groundwater—Purification—Mathematics—Problems, exercises, etc.
I. Title.
TD878.K86 1998
628.1′68—dc21 98-28646
CIP

International Standard Book Number 1-56670-238-0
Library of Congress Card Number 98-28646
Printed in the United States of America 1 2 3 4 5 6 7 8 9 0
Printed on acid-free paper

About the author

Jeff (Jih-Fen) Kuo worked in environmental engineering industries for over 10 years before joining the Department of Civil and Environmental Engineering at California State University, Fullerton, in 1995. He gained his industrial experiences from working at Groundwater Technology, Inc. (now Flour-GTI), Dames and Moore, James M. Montgomery Consulting Engineers (now Montgomery–Watson), Nan-Ya Plastics, and the Los Angeles County Sanitation Districts. His industrial experiences in environmental engineering include design and installation of air strippers, activated carbon adsorbers, flare/catalytic incinerators, and biological systems for groundwater and soil remediation; site assessment and fate analysis of toxics in the environment; RI/FS work for landfills and Superfund sites; design of flanged joints to meet stringent fugitive emission requirements; air emissions from wastewater treatment; and wastewater treatment. Areas of research in environmental engineering include dechlorination of halogenated aromatics by ultrasound, fines/bacteria migration through porous media, biodegradability of bitumen, surface properties of composite mineral oxides, kinetics of activated carbon adsorption, wastewater filtration, THM formation potential of ion exchange resins, and UV disinfection.

He received a B.S. degree in chemical engineering from National Taiwan University, an M.S. degree in chemical engineering from the University of Wyoming, an M.S. in petroleum engineering, and an M.S. and a Ph.D. in Environmental Engineering from the University of Southern California. He is a professional civil, mechanical, and chemical engineer registered in California.

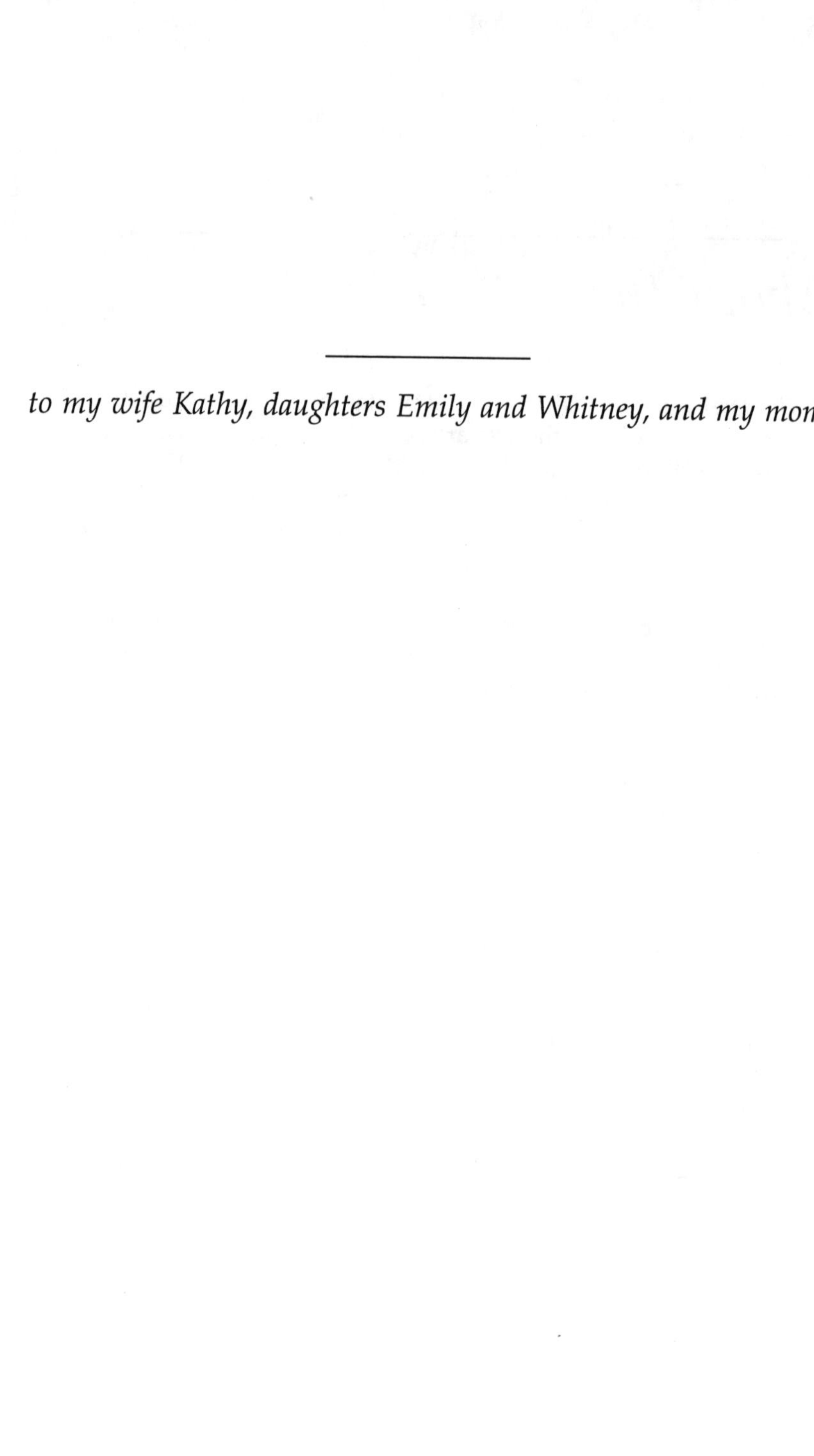

to my wife Kathy, daughters Emily and Whitney, and my mom

Contents

Chapter I Introduction 1
- I.1 Background and Objectives 1
- I.2 Organization 2
- I.3 How to Use this Book 3

Chapter II Site Characterization and Remedial Investigation 5
- II.0 Introduction 5
- II.1 Determination of the Extent of Contamination 7
  - II.1.1 Mass and Concentration Relationship 7
  - II.1.2 Amount of Soil from Tank Removal or Excavation of Contaminated Area 11
  - II.1.3 Amount of Contaminated Soil in the Vadose Zone 15
  - II.1.4 Mass Fractiona and Mole Fraction of Components in Gasoline 17
  - II.1.5 Height of the Capillary Fringe 20
  - II.1.6 Estimating the Mass and Volume of the Free-Floating Product 22
  - II.1.7 Determination of the Extent of Contamination — A Comprehensive Example Calculation 24
- II.2 Soil Borings and Groundwater Monitoring Wells 28
  - II.2.1 Amount of Cuttings from Soil Boring 28
  - II.2.2 Amount of Packing Materials and/or Bentonite Seal29
  - II.2.3 Well Volume for Groundwater Sampling 31
- II.3 Mass of Contaminants Present in Different Phases 32
  - II.3.1 Equilibrium Between Free Product and Vapor 33
  - II.3.2 Liquid–Vapor Equilibrium 36
  - II.3.3 Solid–Liquid Equilibrium 42
  - II.3.4 Solid–Liquid–Vapor Equilibrium 46
  - II.3.5 Partition of Contaminants in Different Phases 48

Chapter III Plume Migration in Groundwater and Soil 57
- III.1 Groundwater Movement 58
  - III.1.1 Darcy's Law 58

III.1.2 Darcy's Velocity vs. Seepage Velocity 59
III.1.3 Intrinsic Permeability vs. Hydraulic Conductivity 61
III.1.4 Transmissivity, Specific Yield, and Storativity 64
III.1.5 Determine Groundwater Flow Gradient and Flow Direction 66
III.2 Groundwater Pumping 67
III.2.1 Steady-State Flow in a Confined Aquifer 67
III.2.2 Steady-State Flow in an Unconfined Aquifer 71
III.3 Aquifer Test 74
III.3.1 Theis Method 75
III.3.2 Cooper–Jacob Straight-Line Method 77
III.3.3 Distance–Drawdown Method 79
III.4 Migration Velocity of the Dissolved Plume 81
III.4.1 The Advection–Dispersion Equation 81
III.4.2 Diffusivity and Dispersion Coefficient 82
III.4.3 Retardation Factor for Migration in Groundwater 88
III.4.4 Migration of the Dissolved Plume 90
III.5 Contaminant Transport in the Vadose Zone 94
III.5.1 Liquid Movement in the Vadose Zone 94
III.5.2 Gaseous Diffusion in the Vadose Zone 95
III.5.3 Retardation Factor for Vapor Migration in the Vadose Zone 98

Chapter IV Mass Balance Concept and Reactor Design 103
IV.1 Mass Balance Concept 104
IV.2 Chemical Kinetics 107
IV.2.1 Rate Equations 107
IV.2.2 Half-Life 110
IV.3 Types of Reactors 112
IV.3.1 Batch Reactors 113
IV.3.2 CFSTRs 117
IV.3.3 PFRs 119
IV.4 Sizing the Reactors 122
IV.5 Reactor Configurations 124
IV.5.1 Reactors in Series 125
IV.5.2 Reactors in Parallel 131

Chapter V Vadose Zone Soil Remediation 139
V.1 Soil Vapor Extraction 139
V.1.1 Introduction 139
V.1.2 Expected Vapor Concentration 140
V.1.3 Radius of Influence and Pressure Profile 148
V.1.4 Vapor Flow Rates 151
V.1.5 Contaminant Removal Rate 155

V.1.6 Cleanup Time ... 158
V.1.7 Effect of Temperature on Soil Venting ... 164
V.1.8 Number of Vapor Extraction Wells ... 165
V.1.9 Sizing of Vacuum Pump (Blower) ... 166
V.2 Soil Bioremediation ... 168
V.2.1 Description of the Soil Bioremediation Process ... 168
V.2.2 Moisture Requirement ... 168
V.2.3 Nutrient Requirements ... 170
V.2.4 Oxygen Requirement ... 172
V.3 Soil Washing/Solvent Extraction/Soil Flushing ... 174
V.3.1 Description of the Soil Washing Process ... 174
V.4 Low-Temperature Heating (Desorption) ... 178
V.4.1 Description of the Low-Temperature Heating (Desorption) Process ... 178
V.4.2 Design of the Low-Temperature Heating (Desorption) Process ... 178

Chapter VI Groundwater Remediation ... 183
VI.1 Hydraulic Control (Groundwater Extraction) ... 183
VI.1.1 Cone of Depression ... 184
VI.1.2 Capture Zone Analysis ... 189
VI.2 Above-Ground Groundwater Treatment Systems ... 198
VI.2.1 Activated Carbon Adsorption ... 198
VI.2.2 Air Stripping ... 205
VI.2.3 Advanced Oxidation Process ... 211
VI.2.4 Metal Removal by Precipitation ... 213
VI.2.5 Biological Treatment ... 214
VI.3 In Situ Groundwater Remediation ... 217
VI.3.1 In Situ Bioremediation ... 217
VI.3.2 Air Sparging ... 222

Chapter VII VOC-Laden Air Treatment ... 229
VII.1 Activated Carbon Adsorption ... 229
VII.1.1 Adsorption Isotherm and Adsorption Capacity ... 230
VII.1.2 Cross-Sectional Area and Height of GAC Adsorbers ... 233
VII.1.3 Contaminant Removal Rate by the Activated Carbon Adsorber ... 234
VII.1.4 Change-Out (or Regeneration) Frequency ... 236
VII.1.5 Amount of Carbon Required (On-Site Regeneration) ... 237
VII.2 Thermal Oxidation ... 237
VII.2.1 Air Flow Rate vs. Temperature ... 238
VII.2.2 Heating Values of an Air Stream ... 239
VII.2.3 Dilution Air ... 241

VII.2.4 Auxiliary Air to Supply Oxygen 243
VII.2.5 Supplementary Fuel Requirements 245
VII.2.6 Volume of Combustion Chamber 246
VII.3 Catalytic Incineration 248
VII.3.1 Dilution Air 248
VII.3.2 Supplementary Heat Requirements 249
VII.3.3 Volume of the Catalyst Bed 250
VII.4 Internal Combustion Engines 251
VII.4.1 Sizing Criteria/Application Rates 252
VII.5 Soil Beds/Biofilters 252
VII.5.1 Design Criteria 252

Index 255

Preface

The focus of the hazardous waste management business has switched in recent years from litigation and site assessment to remediation. Site restoration usually proceeds through several phases and requires a concerted, multidisciplinary effort. Thus, remediation specialists have a variety of backgrounds, including geology, hydrology, chemistry, microbiology, meteorology, toxicology, and epidemiology as well as chemical, mechanical, electrical, civil, and environmental engineering. Because of differences in the formal education of these professionals, their ability to perform or review remediation design calculations varies considerably. For some, performing accurate design calculations can become a seemingly insurmountable task.

Most, if not all, of the books dealing with site remediation provide only descriptive information on treatment technologies, and none, in my opinion, provide helpful guidance on illustrations of design calculations. This book was written to address the current needs of practicing engineers, scientists, and legal experts who are employed by industry, consulting companies, law firms, and regulatory agencies as well as university seniors and graduate students in the field of soil and groundwater remediation. It provides practical and relevant working information, derived from the literature and from my own hands-on experiences in consulting and teaching in this field. I sincerely hope that this book becomes a useful tool for the professionals and students working in site remediation. Your comments and suggestions are always welcome, and my e-mail address is jkuo@fullerton.edu.

Finally, I would like to take this opportunity to thank Tom Hashman and Ziad El Jack of the Sanitation Districts of Los Angeles County for reviewing the manuscript and providing valuable comments.

chapter one

Introduction

I.1 Background and objectives

The hazardous waste management business has steadily increased since the early 1980s as public concern led to a vast range of new environmental regulations. During much of this period, a substantial amount of time and expense has been devoted to studying contaminated sites, with much of the expense dedicated to litigation to determine the financially responsible parties. However, the focus has switched in recent years from litigation and site assessment to remediation. Site restoration usually proceeds through several phases and requires a concerted, multidisciplinary effort. Thus, remediation specialists have a variety of backgrounds, including geology, hydrology, chemistry, microbiology, meteorology, toxicology, and epidemiology as well as chemical, mechanical, electrical, civil, and environmental engineering. Because of differences in the formal education of these professionals, their ability to perform or review remediation design calculations varies considerably. For some, performing accurate design calculations can become a seemingly insurmountable task.

The absence of uniformly trained specialists is exacerbated by the continuously evolving remediation technology. For instance, remediation technologies such as soil venting and bioremediation are now generally acceptable to environmental professionals, while just a few years ago they were the subject mainly of research articles. While up-to-date design information is sporadically published in the literature, it is usually theoretical in nature and illustrative applications are rarely given. Most, if not all, of the books dealing with hazardous waste management and site remediation provide only descriptive information on treatment technologies, and none, in this author's opinion, provide helpful guidance on illustrations of design calculations.

Without the proper education, environmental professionals can exert themselves, needlessly reinventing the wheel, so to speak, and err in their design calculations. This book was written to address the current needs of practicing engineers, scientists, and legal experts who are employed by industry, consulting companies, law firms, and regulatory agencies in the field of soil and groundwater remediation. It covers important aspects of the major design calculations used in this field and also provides practical and relevant working information derived from the literature and the author's own experience. Realistic examples are used liberally to illustrate the application of the design calculations. This book can also serve as a supplementary textbook or reference book for university seniors and graduate students who would like to have an overview of remediation design calculations.

I.2 Organization

The book is divided into the following chapters:

Chapter two: Site Assessment and Remedial Investigation. This chapter illustrates engineering calculations needed during site assessment and remedial investigation. It begins with simple calculations for estimating the amount of contaminated soil excavated and that left in the vadose zone and size of the contaminated plume in the aquifer. This chapter also describes necessary calculations to determine partitioning of contaminant mass in the different phases, which is critical for remediation design.

Chapter three: Groundwater Movement and Plume Migration. This chapter illustrates how to estimate the rates of groundwater movement and plume migration. The reader will also learn how to interpret the aquifer test data and estimate the age of a groundwater plume.

Chapter four: Mass Balance Concept and Reactor Design. This chapter first introduces the mass balance concept, followed by reaction kinetics, as well as types, configuration, and sizing of reactors. The reader will learn how to determine the rate constant, removal efficiency, optimal arrangement of reactors, required residence time, and reactor size for one's specific applications.

Chapter five: Vadose Zone Soil Remediation. This chapter provides important design calculations for commonly used in situ or above-ground soil remediation techniques, such as soil vapor extraction, soil washing, and soil bioremediation. Taking soil venting as an example, the book will guide the reader through design calculations for radius of influence, well spacing, air flow rate, extracted contaminant concentrations, effect of temperature on vapor flow, cleanup time, and sizing of vacuum blowers.

Chapter six: Groundwater Remediation. This chapter starts with design calculations for capture zone and optimal well spacing. The rest of the chapter focuses on design calculations for commonly used in situ or ex situ groundwater remediation techniques, including bioremediation, air sparging, air stripping, advanced oxidation process, and activated carbon adsorption.

Chapter seven: VOC-Laden Air Treatment. Remediation of contaminated soil and groundwater often results in transferring organic contaminants into the air phase. Development and implementation of an air emission control strategy are an integral part of the overall remediation program. This chapter illustrates design calculations for commonly used off-gas treatment technologies such as activated carbon adsorption, direct incineration, catalytic incineration, IC engines, and biofiltration.

I.3 How to use this book

The book is constructed to provide a comprehensive coverage of commonly used soil and groundwater remediation technologies. It is written in a cookbook style and user-friendly format. Both SI and U.S. customary units are used throughout the book, and unit conversions are frequently given. Examples are given following the design equations. Some of the examples are provided to illustrate important design concepts. One of the best ways to use the book is to glance through the entire book first, by reading the text and skimming the problem statement and discussion only, and revisit the specific topics in detail later when related design calculations are to be made.

chapter two

Site characterization and remedial investigation

II.0 Introduction

The initial step, often the most critical one, of a typical soil and/or groundwater remediation project is to determine the extent of contamination. It is often accomplished by site characterization and remedial investigation (RI).

Site characterization is to determine the conditions on and beneath a site that are pertinent to hazardous waste management. When site remediation is deemed necessary, RI will be employed. RI activities consist of site characterization and additional data collection. The additional data are necessary for control of plume migration and selection of remedial alternatives. The common questions to be answered by the RI activities are, "Where is the contaminant plume? What is in the plume? How big is the plume? How long has it been there? Where is it going? How fast will it go?"

Subsurface contamination from spills and leaky underground storage tanks (USTs) creates environmental problems that usually require corrective actions. The contaminants may be present in one or a combination of the following locations and phases:

Vadose zone

- Vapors in the void
- Free product in the void
- Dissolved in soil moisture
- Adsorbed onto the soil matrix
- Floating on top of the capillary fringe (for nonaqueous phase liquids [NAPLs])

Groundwater

- Dissolved in the groundwater

- Adsorbed onto the aquifer material
- Sitting on top of the bedrock (for dense nonaqueous phase liquids [DNAPLs])

Common RI activities include:

1. Removal of contamination source(s) such as leaky USTs
2. Installation of soil borings
3. Installation of groundwater monitoring wells
4. Soil sample collection and analysis
5. Groundwater sample collection and analysis
6. Aquifer testing

Through these activities, the following data are collected:

1. Types of contaminants present in soil and groundwater
2. Concentrations of contaminants in the collected samples
3. Vertical and areal extents of contaminant plumes in soil and groundwater
4. Vertical and areal extents of free-floating product or the DNAPLs
5. Soil characteristics including the types of soil, density, moisture content, etc.
6. Groundwater elevations
7. Drawdown data collected from aquifer tests

Using these collected data, engineering calculations are then performed to assist site remediation. Common engineering calculations include:

1. Mass and volume of soil excavated during tank removal
2. Mass and volume of contaminated soil left in the vadose zone
3. Mass of contaminants in the vadose zone
4. Mass and volume of the free-floating product
5. Volume of contaminated groundwater
6. Mass of contaminants in the aquifer
7. Groundwater flow gradient and direction
8. Hydraulic conductivity of the aquifer

This chapter describes all the above-needed engineering calculations, except the last two, which will be covered in Chapter 3. Discussions will also be presented concerning the calculations related to site activities, including cuttings from soil boring and purge water from groundwater sampling. The last part of the chapter describes the "partitioning" of contaminants in different phases. Understanding the partitioning phenomena of the contaminants is critical for studying the fate and transport of contaminants in the subsurface and for selection of remedial alternatives.

II.1 Determination of the extent of contamination

II.1.1 Mass and concentration relationship

As mentioned earlier, contaminants may exist in different phases. In environmental engineering applications, people commonly express contaminant concentrations in parts per million (ppm), parts per billion (ppb), or parts per trillion (ppt).

Although these concentration units are commonly used, some people may not realize that "one ppm," for example, does not mean the same for liquid, solid, and air phases. In the liquid and solid phases, the ppm unit is on a mass per mass basis. One ppm stands for one part mass of a compound in one million parts mass of the media containing it. Soil contaminated with one ppm benzene means that every gram of soil contains one microgram of benzene, i.e., 10^{-6} g benzene per gram of soil, or 1 mg benzene per kilogram of soil (1 mg/kg).

For the liquid phase, one ppm of benzene means 1 μg of benzene dissolved in 1 g of water, or 1 mg benzene per kilogram water. Since it is usually more convenient to measure the liquid volume than its mass, and 1 kg of water has a volume of approximately 1 L under ambient conditions, people commonly use "1 ppm" for "1 mg/L compound concentration in liquid."

For the vapor phase, the story is totally different. One ppm by volume (ppmV) is on a volume per volume basis. One ppmV of benzene in the air means one part volume of benzene in one million parts volume of air space. To convert the ppmV into mass concentration units, which is often needed in remediation work, we can use the following formula:

$$
\begin{aligned}
1\ \text{ppmV} &= \frac{\text{MW}}{22.4}[\text{mg/m}^3] \quad \text{at } 0°\text{C} \\
&= \frac{\text{MW}}{24.05}[\text{mg/m}^3] \quad \text{at } 20°\text{C} \qquad \text{[Eq. II.1.1]} \\
&= \frac{\text{MW}}{24.5}[\text{mg/m}^3] \quad \text{at } 25°\text{C}
\end{aligned}
$$

or

$$
\begin{aligned}
1\ \text{ppmV} &= \frac{\text{MW}}{359}\times 10^{-6}[\text{lb/ft}^3] \quad \text{at } 32°\text{F} \\
&= \frac{\text{MW}}{385}\times 10^{-6}[\text{lb/ft}^3] \quad \text{at } 68°\text{F} \qquad \text{[Eq.II.1.2]} \\
&= \frac{\text{MW}}{392}\times 10^{-6}[\text{lb/ft}^3] \quad \text{at } 77°\text{F}
\end{aligned}
$$

where MW is the molecular weight of the compound, and the number in the denominator of each equation above is the molar volume of an ideal gas at that temperature. For example, the volume of an ideal gas is 22.4 L per gram-mole at 0°C, or 359 ft^3 per pound-mole at 32°F.

Let us determine the conversion factors between ppmV and mg/m^3 or lb/ft^3, using benzene (C_6H_6) as an example. The molecular weight of benzene is 78, therefore 1 ppmV of benzene is the same as

$$
\begin{aligned}
1 \text{ ppmV benzene} &= \frac{78}{24.05} = 3.24 \text{ mg/m}^3 \quad \text{at } 20°\text{C} \\
&= \frac{78}{24.5} = 3.18 \text{ mg/m}^3 \quad \text{at } 25°\text{C} \qquad \text{[Eq. II.1.3]} \\
&= \frac{78}{392} \times 10^{-6} = 0.199 \times 10^{-6} \text{ lb/ft}^3 \quad \text{at } 77°\text{F } (25°\text{C})
\end{aligned}
$$

From this practice, we learn that the conversion factors are different among compounds because of the differences in molecular weight. In addition, the conversion factor for a compound is temperature dependent because its molar volume varies with temperature.

In remediation design, it is often necessary to determine the mass of a contaminant present in a medium. It can be found from the contaminant concentration and the amount of the medium containing the contaminant. The procedure for such calculations is simple but slightly different for the liquid, soil, and air phases. The differences mainly come from the concentration units.

Let us start with the simplest case that a liquid is polluted with a dissolved contaminant. Dissolved contaminant concentration in the liquid (C) is often expressed in mass of contaminant/volume of liquid, such as milligrams per liter, therefore, mass of the contaminant in the liquid can be obtained by multiplying the concentration by the volume of liquid (V_l):

$$
\begin{aligned}
&\text{Mass of contaminant in liquid} = \\
&(\text{liquid volume})(\text{liquid concentration}) = (V_l)(C) \qquad \text{[Eq. II.1.4]}
\end{aligned}
$$

Contaminant concentration on a soil surface (X) is often expressed in mass of contaminant/mass of soil, such as milligrams per kilogram; therefore, the mass of contaminants can be obtained by multiplying the concentration with the mass of soil (M_s). Mass of soil, in turn, is the multiplication product of volume of soil (V_s) and bulk density of soil (ρ_b):

$$
\begin{aligned}
\text{Mass of contaminant in soil} &= (X)(M_S) \qquad \text{[Eq. II.1.5]} \\
&= (X)[(V_S)(\rho_b)]
\end{aligned}
$$

Contaminant concentration in air (G) is often expressed in vol/vol such as ppmV or in mass/vol such as mg/m³. In calculation of mass, we need to convert the concentration into the mass/vol basis first using Eq. II.1.2. Mass of the contaminant in air can then be obtained by multiplying the concentration with the volume of air (V_a):

$$\text{Mass of contaminant in air} = (\text{air volume})(\text{concentration in mass/vol}) = (V_a)(G) \qquad \text{[Eq. II.1.6]}$$

Example II.1.1A Mass and concentration relationship

Which of the following media contains the largest amount of xylene?

a. 1 million gallons of water containing 10 ppm of xylene
b. 100 cubic yards of soil (bulk density = 1.8 g/cm³) with 10 ppm of xylene
c. An empty warehouse (200′ × 50′ × 20′) with 10 ppmV xylene in air

Solution:

a. Mass of contaminant in liquid = (liquid volume)(liquid concentration) = (1,000,000 gallon)(3.785 L/gallon)(10 mg/L) = 3.79×10^7 mg
b. Mass of contaminant in soil = (soil volume)(density)(soil concentration)

$$= [(100\ \text{yd}^3)(27\ \text{ft}^3/\text{yd}^3)(30.48\ \text{cm/ft})^3]\ [(1.8\ \text{g/cm}^3(\text{kg}/1000\text{g})](10\ \text{mg/kg})$$

$$= 1.37 \times 10^6\ \text{mg}$$

c. Molecular weight of xylene [$C_6H_4(CH_3)_2$]

$$= (12)(6) + (1)(4) + (12 + 1 \times 3)(2) = 106\ \text{g/mole}$$

$$10\ \text{ppmV} = (10)(\text{MW of xylene}/24.05)\ \text{mg/m}^3$$

$$= (10)(106/24.05)\ \text{mg/m}^3 = 44.07\ \text{mg/m}^3$$

Mass of contaminant in air = (air volume)(vapor concentration)

$$= [(200 \times 50 \times 20\ \text{ft}^3)(0.3048\ \text{m/ft})^3](44.07\ \text{mg/m}^3)$$

$$= 2.5 \times 10^5\ \text{mg}$$

The water contains the largest amount of xylene.

Example II.1.1B *Mass and concentration relationship*

If a person drinks 2 L of water containing 1 ppb of benzene and inhales 20 m^3 of air containing 10 ppbV of benzene a day, which system (ingestion or inhalation) is exposed to more benzene?

Solution:

a. Benzene ingested daily:

$$(2\ L)(10^{-3}\ mg/L) = 2 \times 10^{-3}\ mg$$

b. Molecular weight of benzene (C_6H_6) = (12)(6) +(1)(6) = 78 g/mole

$$10\ ppbV = (10 \times 10^{-3})(78/24.05)\ mg/m^3 = 0.0324\ mg/m^3$$

Benzene inhaled daily:

$$(20\ m^3)(0.0324\ mg/m^3) = 0.65\ mg$$

The inhalation system is exposed to more benzene.

Example II.1.1C *Mass and concentration relationship*

A glass bottle containing 900 mL of methylene chloride (CH_2Cl_2, specific gravity = 1.335) was accidentally left uncapped over a weekend in a poorly ventilated room (5 m × 6 m × 3.6 m). On the following Monday it was found that two thirds of methylene chloride had volatilized. For a worst-case scenario, would the concentration in the room air exceed the permissible exposure limit (PEL) of 100 ppmV?

Solution:

a. Mass of methylene chloride volatilized = (liquid volume)(density) $[(2/3)(900\ mL)(1\ mL/cm^3)](1.335\ g/cm^3) = 801\ g = 8.01 \times 10^5\ mg$

b. Vapor concentration in mass/vol = (mass) ÷ (volume)

$$(8.01 \times 10^5\ mg) \div [(5\ m)(6\ m)(3.6\ m)] = 7417\ mg/m^3$$

c. Molecular weight of methylene chloride [CH_2Cl_2] = (12) + (1)(2) + (35.5)(2) = 85 g/mole

$$1\ ppmV = (85/24.05)\ mg/m^3 = 3.53\ mg/m^3$$

Vapor concentration in vol/vol = 7417 mg/m^3 ÷ [3.53 (mg/m^3)/ppmV] = 2100 ppmV

It would exceed the PEL.

Example II.1.1D *Mass and concentration relationship*

A child went into a site and played with dirt contaminated with benzene. During his stay at the site he inhaled 2 m^3 of air containing 10 ppbV of benzene and ingested a mouthful (~5 cm^3) of soil containing 3 mg/kg of benzene. Which system (ingestion or inhalation) is exposed to more benzene? Assume the bulk density of soil is 1.8 g/cm^3.

Solution:

a. 10 ppbV of benzene = $(10 \times 10^{-3})(78/24.05)$ mg/m^3 = 0.0324 mg/m^3

Mass of benzene inhaled = (air volume)(vapor concentration) = (2 m^3)(0.0324 mg/m^3) = 0.065 mg

b. Benzene ingested = (volume of soil)(density of soil)(soil concentration)

[(5 cm^3)(1.8 g/cm^3)(1 kg/1000 g)](3 mg/kg) = 0.027 mg

The inhalation system is exposed to more benzene.

II.1.2 *Amount of soil from tank removal or excavation of contaminated area*

Removal of USTs typically involves soil excavation. If the excavated soil is clean (i.e., free of contaminants or below the permissible levels), it may be reused as backfill materials or disposed of in a sanitary landfill. On the other hand, if it is contaminated, it needs to be treated or disposed of in a hazardous waste landfill. For either case, a good estimate of soil volume and/or mass is necessary.

The excavated soil is usually stored on site first as stockpiles. The amount of excavated soil from tank removal can be determined from measurement of the volumes of the stockpiles. However, the shapes of these piles are irregular, and this makes the measurement more difficult. An easier and more accurate alternative is

Step 1: Measure the dimensions of the tank pit.
Step 2: Calculate the volume of the tank pit from the measured dimensions.

Step 3: Determine the number and volumes of the USTs removed.
Step 4: Subtract the total volume of the USTs from the volume of the tank pit.
Step 5: Multiply the value from Step 4 with a soil fluffy factor.

Information needed for this calculation

- Dimensions of the tank pit (from field measurement)
- Number and volumes of the USTs removed (from drawings or field measurement)
- Density of soil (from measurement or estimate)
- Soil fluffy factor (from estimate)

Example II.1.2A *Determine the mass and volume of soil excavated from a tank pit*

Two 5000-gal USTs and one 4000-gal UST were removed. The excavation resulted in a tank pit of 50′ × 24′ × 18′. The excavated soil was stockpiled on-site. The bulk density of soil in situ (before excavation) is 1.8 g/cm^3, and bulk density of soil in the stockpiles is 1.64 g/cm^3. Estimate the mass and volume of the excavated soil.

Solution:

Volume of the tank pit = (50′)(24′)(18′) = 21,600 ft^3

Total volume of the USTs = (2)(5000) + (1)(4000) = 14,000 gallons

= (14,000 gallon)(ft^3/7.48 gallon) = 1872 ft^3

Volume of soil in the tank pit before removal = (volume of tank pit) – (volume of USTs)

= 21,600 – 1972 = 19,728 ft^3

Volume of soil excavated (in the stockpile) = (volume of soil in the tank pit) × (fluffy factor)

= (19,728)(1.10) = 21,700 ft^3 = (21,700 ft^3)[yd^3/27 ft^3] = 804 yd^3

Mass of soil excavated = (volume of the soil in the tank pit)(bulk density of soil in situ) = (volume of the soil in the stockpile)(bulk density of soil in the stockpile)

Soil density in situ = 1.8 g/cm^3 = (1.8 g/cm^3)[(62.4 lb/ft^3)/(1g/cm^3)]

= 112 lb/ft^3

Soil density in stockpiles = (1.64)(62.4) = 102 lb/ft^3

Mass of soil excavated = (19,728 ft^3)(112 lb/ft^3) = 2,210,000 lb = 1100 tons

or = (21,700 ft^3)(102 lb/ft^3) = 2,210,000 lb = 1100 tons

Discussion. The fluffy factor of 1.10 is to take into account the expansion of soil after being excavated from subsurface. The in situ soil is usually more compacted. A fluffy factor of 1.10 means the volume of soil increases by 10% from in situ to above ground. On the other hand, the bulk density of soil in the stockpiles would be lower than that of in situ soil as the result of expansion after excavation.

Example II.1.2B   Mass and concentration relationship of excavated soil

A leaky 4.5-m^3 underground storage tank was removed. The excavation resulted in a tank pit of 4 m × 4 m × 5 m (L × W × H), and the excavated soil was stockpiled on site. Three samples were taken from the pile and the TPH concentrations were determined to be <100, 1500, and 2000 ppm. What is the amount of TPH in the pile? Express your answers in both kilograms and liters.

Solution:

Volume of the tank pit = (4)(4)(5) = 80 m^3

Volume of soil in the tank pit before removal = (volume of tank pit) – (volume of USTs)

= 80 – 4.5 = 75.5 m^3

Average TPH concentration = (100 + 1500 + 2000)/3

= 1200 ppm = 1200 mg/kg

Mass of TPH in soil = [(75.5 m^3)(1800 kg/m^3)](1200 mg/kg)

= 1.63×10^8 mg = 163 kg

Volume of TPH in soil = (mass of TPH)/(density of TPH)

= (163 kg)/(0.8 kg/L) = 203.8 L = 53.9 gallons

Discussion

1. The bulk density of soil was assumed to be 1800 kg/m^3 (i.e., 1.8 g/cm^3), and the density of total petroleum hydrocarbon (TPH) was assumed to be 0.8 kg/L (i.e., 0.8 g/cm^3).
2. The TPH concentration for one of the three samples is below the detection limit (<100 ppm). Four methods are common for dealing with values below the detection limit: (1) use the detection limit as the value, (2) use half of the detection limit, (3) use zero, and (4) select a value based on a statistical approach (especially when multiple samples are taken and a few of them are below the detection limit). In this solution, a conservative approach was taken by using the detection limit as the concentration.

Example II.1.2C *Mass and concentration relationship of excavated soil*

A leaky 1000-gal underground storage tank was removed. The excavation resulted in a tank pit of 12′ × 12′ × 15′ (L × W × H), and the excavated soil was stockpiled on site. Five samples were taken from the pile and analyzed for TPH using EPA method 8015. Based on the laboratory results, an engineer at CSUF Consulting Company estimated that there were approximately 50 gal of gasoline present in the soil pile. One of the five TPH values in the report was illegible, and the others were <100, 1000, 2000, and 3000 ppm, respectively. What is the missing value?

Solution:

Average TPH concentration = $(x + 100 + 1000 + 2000 + 3000)/5$

Mass of contaminated soil = $[(12)(12)(15) - (1000/7.48)](112)$

$$= 227{,}000 \text{ lb} = 103{,}000 \text{ kg}$$

Mass of TPH in soil = (volume of gasoline)(density of gasoline)

$$= [(50 \text{ gal})(\text{ft}^3/7.48 \text{ gal})](50 \text{ lb/ft}^3)(\text{kg}/2.2 \text{ lb}) = 151.9 \text{ kg}$$

= (contaminant concentration)(mass of contaminated soil)

$$= [(x + 100 + 1000 + 2000 + 3000)/5 \text{ mg/kg}](103{,}000 \text{ kg})(\text{kg}/10^6 \text{ mg})$$

x = the unknown TPH concentration = 1264 ppm

II.1.3 Amount of contaminated soil in the vadose zone

Chemicals that leak from USTs might move beyond the tank pit. If subsurface contamination is suspected, soil borings are often drilled to assess the extent of contamination in the vadose zone. Soil boring samples are then taken at a fixed interval, e.g., every 5 or 10 ft, and analyzed for soil properties. Selected samples are submitted to laboratories and analyzed for contaminant concentrations. From these data, a contaminant fence diagram is often developed to delineate the extent of the contaminant plume.

When selecting remedial alternatives, an engineer needs to know the location of the plume, types of subsurface soil, types of contaminants, mass and/or volume of the contaminated soil, and mass of contaminants. If the location of the plume is shallow (not deep from the ground level surface) and the amount of contaminated soil is not extensive, excavation coupled with above-ground treatment may be a viable option. On the other hand, in situ remediation alternatives such as soil venting would be more favorable if the volume of the contaminated soil is large and deep. Therefore, a good estimate of the amount of contaminated soil left in the vadose zone is important for remediation design. This section describes the methodology for such calculations.

As mentioned, a fence diagram is often drawn to illustrate the vertical and areal extents of the plume. Based on the information from the diagram, the following procedure can be used to determine the amount of contaminated soil in the vadose zone:

Step 1: Determine the area of contaminated plume at each sampling depth, A_i.
Step 2: Determine the thickness interval for each area calculated above, h_i.
Step 3: Determine the volume of the contaminated soil, V_s, using the following formula:

$$V_S = \sum_i A_i h_i \qquad \text{[Eq. II.1.7]}$$

Step 4: Determine the mass of the contaminated soil, M_s, by multiplying V_s by the density of soil, ρ_b, as

$$M_S = \rho_b \times V_S \qquad \text{[Eq. II.1.8]}$$

Information needed for this calculation

- The areal and vertical extent of the plume, A_i and h_i
- Bulk density of soil, ρ_b

To determine the mass and volume of contaminated water contained in a groundwater plume, the following procedure should be followed:

Step 1: Use Eq. II.1.7 to determine the size of the plume.
Step 2: Multiply the volume from Step 1 by aquifer porosity to obtain the volume of groundwater.
Step 3: Multiply the volume from Step 2 by water density to obtain the mass of contaminated water.

Example II.1.3A Determine the amount of contaminated soil in the vadose zone

For the project described in Example II.1.2A, after the USTs were removed, five soil borings were installed. Soil samples were taken every 5 ft below ground surface (bgs). Based on the laboratory analytical results and subsurface geology, the area of the plume at each soil sampling interval was determined as follows:

Depth (ft bgs)	Area of the plume (ft^2)
15	0
20	350
25	420
30	560
35	810
40	0

Determine the volume and mass of contaminated soil left in the vadose zone.

Strategy. The soil samples were taken and analyzed every 5 ft; therefore, each plume area represents the same depth interval. The sample taken at 20-ft depth represents the 5-ft interval from 17.5 to 22.5 ft (the midpoint of the first two consecutive intervals to the midpoint of the next two consecutive intervals), the sample at 25-ft depth represents the 5-ft interval from 22.5 ft to 27.5 ft, and so on.

Solution:

Thickness interval for each area is the same at 5 ft

Volume of the contaminated soil (using Eq. II.1.7)

$$= (5)(350) + (5)(420) + (5)(560) + (5)(810) = 10{,}700 \text{ ft}^3 = 396 \text{ yd}^3$$

$$\text{or } (22.5 - 17.5)(350) + (27.5 - 22.5)(420) + (32.5 - 27.5)(560)$$

$$+ (37.5 - 32.5)(810) = 10{,}700 \text{ ft}^3$$

Mass of the contaminated soil (using Eq. II.1.8)

$$= (10{,}700 \text{ ft}^3)(112 \text{ lb/ft}^3) = 1{,}198{,}400 \text{ lb} = 599 \text{ tons}$$

Example II.1.3B *Determine the amount of contaminated soil in the vadose zone*

For the project described in Example II.1.2A, after the USTs were removed, five soil borings were installed. Soil samples were taken every 5 ft bgs. However, not all the samples were analyzed because of budget constraints. Based on the laboratory analytical results and subsurface geology, the area of the plume at a few depths were determined as follows:

Depth (ft bgs)	Area of the plume (ft^2)
15	0
20	350
25	420
35	810
40	0

Determine the volume and mass of the contaminated soil left in the vadose zone.

Strategy. The depth intervals given are not the same as before; therefore, each plume area represents a different depth interval. For example, the sample taken at 25-ft depth represents a 7.5-ft interval, from 22.5 ft to 30 ft.

Solution:

Volume of the contaminated soil (using Eq. II.1.7)

$$= (5)(350) + (7.5)(420) + (7.5)(810) = 10{,}915 \text{ ft}^3 = 406 \text{ yd}^3$$

$$\text{or} = (22.5 - 17.5)(350) + (30 - 22.5)(420) + (37.5 - 30)(810) = 10{,}915 \text{ ft}^3$$

Mass of the contaminated soil (using Eq. II.1.8)

$$= (10{,}975 \text{ ft}^3)(112 \text{ lb/ft}^3) = 1{,}229{,}200 \text{ lb} = 615 \text{ tons}$$

II.1.4 *Mass fraction and mole fraction of components in gasoline*

Gasoline is a common contaminant found in the subsurface as a result of leaky USTs. Gasoline itself is a mixture of various hydrocarbons, and it may

Table II.1.A Some Physical Properties of BTEX

	Formula	M.W.	Water solubility (mg/L)	Vapor pressure (mmHg)
Benzene	C_6H_6	78	1780 @ 25°C	95 @ 25°C
Toluene	$C_6H_5(CH_3)$	92	515 @ 20°C	22 @ 20°C
Ethylbenzene	$C_6H_5(C_2H_5)$	106	152 @ 20°C	7 @ 20°C
Xylenes	$C_6H_4(CH_3)_2$	106	198 @ 20°C	10 @ 20°C

From U.S. EPA, *CERCLA Site Discharges to POTWs Treatability Manual,* EPA 540/2-90-007, Office of Water, U.S. EPA, Washington, DC, 1990.

From LaGrega, M.D. , Buckingham, P.L., and Evans, J.C., *Hazardous Waste Management,* McGraw-Hill, New York, 1994. With permission.

contain more than 200 different compounds. Some of them are lighter and more volatile than the others (lighter ends vs. heavier ends). Gasoline in soil samples is usually measured by EPA method 8015 as total petroleum hydrocarbon (TPH), using gas chromatography (GC). It can also be measured by EPA method 418.1, using infrared (IR), which is considered more suitable for heavier-end hydrocarbons. Diesel fuel is often measured by "modified" EPA method 8015 that takes into account the abundance of heavier ends in diesel fuel as compared to gasoline. Some of the gasoline constituents are more toxic than the others. Benzene, toluene, ethylbenzene, and xylenes (BTEX) are gasoline constituents of concern because of their toxicity. (Benzene is a known carcinogen.) BTEX compounds are measured by EPA method 8020. To cut down the air pollution, many oil companies have developed so-called "new-formula" gasoline, in which the benzene content is reduced. Some of the important physical properties of BTEX are tabulated in Table II.1.A.

Sometimes, it is necessary to determine the composition, such as mass and mole fractions of important compounds, of the gasoline for the following reasons:

1. *Identification of responsible parties.* At a busy intersection having two or more gasoline stations, the free-floating product found beneath a site may not come from its USTs. Each brand of gasoline usually has its own distinct formula, mainly due to differences in refining processes or in the crude oils. Most oil companies have the capabilities to identify the biomarkers in the gasoline or to determine if the composition of the free-floating product matches their formula.
2. *Determination of health risk.* As mentioned, some gasoline constituents are more toxic than the others, and they should be considered differently in a risk assessment.
3. *Determination of the product age.* Some compounds are more volatile than others. The fraction of volatile constituents in a recently spilled gasoline should be larger than that in an aged spill.

To determine the mass fractions of compounds in gasoline, the following procedure can be used:

Step 1: Determine the mass of TPH and mass of each compound of concern.
Step 2: Determine the mass fraction by dividing the mass of the compound by the mass of TPH.

To determine the mole fractions of compounds in gasoline, the following procedure can be used:

Step 1: Determine the mass of TPH and mass of each compound of concern in contaminated soil.
Step 2: Determine the molecular weight of each compound.
Step 3: Determine the molecular weight of gasoline from the composition and the molecular weights of all constituents. This procedure is tedious, and information may not be readily available. Assuming the molecular weight of gasoline to be 100, which is equivalent to that of heptane (C_7H_{16}), is relatively reasonable.
Step 4: Determine the number of moles of each compound by dividing its mass by its molecular weight.
Step 5: Calculate the mole fraction by dividing the number of moles of each compound with the number of moles of the TPH.

Information needed for this calculation

- Mass of contaminated soil
- Contaminant concentrations
- Molecular weights of the contaminants

Example II.1.4 *Mass and mole fractions of components in gasoline*

Three samples were taken from a soil pile (110 yd^3) and analyzed for TPH (EPA method 8015) and for BTEX (EPA method 8020). The average concentration of TPH is 1000 mg/kg, and those of BTEX are 85, 50, 35, and 40 mg/kg, respectively. Determine the mass and mole fractions of BTEX in the gasoline. The bulk density of the soil is 1.65 g/cm^3.

Solution:

a. Mass of contaminated soil = (volume of soil)(bulk density)

$$= [(110\ \text{yd}^3)(27\ \text{ft}^3/\text{yd}^3)]\{(1.65\ \text{g/cm}^3)[62.4\ \text{lb/ft}^3/(1\ \text{g/cm}^3)]\}$$

$$= 305{,}800\ \text{lb} = 139{,}000\ \text{kg}$$

b. Mass of a contaminant in soil = (soil mass)(contaminant concentration)
Mass of TPH = (139,000 kg)(1000 mg/kg) = 1.39×10^8 mg = 1.39×10^5 g

Mass of benzene = (139,000 kg)(85 mg/kg) = 1.181×10^7 mg

= 1.181×10^4 g

Mass of toluene = (139,000 kg)(50 mg/kg) = 6.950×10^6 mg

= 6.950×10^3 g

Mass of ethylbenzene = (139,000 kg)(35 mg/kg) = 4.865×10^6 mg

= 4.865×10^3 g

Mass of xylenes = (139,000 kg)(40 mg/kg) = 5.560×10^6 mg

= 5.560×10^3 g

c. Mass fraction of a compound = (mass of the compound)/(mass of TPH)

Mass fraction of benzene = $(1.181 \times 10^4)/(1.39 \times 10^5) = 0.085$
Mass fraction of toluene = $(6.95 \times 10^3)/(1.39 \times 10^5) = 0.05$
Mass fraction of ethylbenzene = $(4.865 \times 10^3)/(1.39 \times 10^5) = 0.035$
Mass fraction of xylenes = $(5.56 \times 10^3)/(1.39 \times 10^5) = 0.04$

d. Moles of a compound = (mass of the compound)/(molecular weight of the compound)

Moles of TPH = $(1.39 \times 10^5)/(100) = 1390$ g-mole
Moles of benzene = $(1.181 \times 10^4)/(78) = 151.4$ g-mole
Moles of toluene = $(6.95 \times 10^3)/(92) = 77.5$ g-mole
Moles of ethylbenzene = $(4.865 \times 10^3)/(106) = 45.9$ g-mole
Moles of xylenes = $(5.56 \times 10^3)/(106) = 52.5$ g-mole

e. Mole fraction of a compound = (moles of the compound)/(moles of TPH)

Mole fraction of benzene = (151.4)/(1390) = 0.109
Mole fraction of toluene = (77.5)/(1390) = 0.056
Mole fraction of ethylbenzene = (45.9)/(1390) = 0.033
Mole fraction of xylenes = (52.5)/(1390) = 0.038

Discussion. The mass fraction of each compound can also be determined directly from the ratio of the compound concentration to the TPH concentration. Using benzene as an example, mass fraction of benzene = (85 mg/kg)/(1000 mg/kg) = 0.085 = 8.5%.

II.1.5 Height of the capillary fringe

The capillary fringe (or capillary zone) is a zone immediately above the water table of unconfined aquifers. It extends from the top of the water table due to the capillary rise of water. The capillary fringe often creates complications

Table II.1.B Typical Height of Capillary Fringe

Material	Grain size (mm)[a]	Pore radius (cm)[b]	Capillary rise (cm)	
Coarse gravel		0.4		0.38[b]
Fine gravel	5–2		2.5[a]	
Very coarse sand	2–1		6.5[a]	
Coarse sand	1–0.5	0.05	13.5[a]	3.0[b]
Medium sand	0.5–0.2		24.6[a]	
Fine sand	0.2–0.1	0.02	42.8[a]	7.7[b]
Silt	0.1–0.05	0.001	105.5[a]	150[b]
Silt	0.05–0.02		200[a]	
Clay		0.0005		300[b]

[a] Reid, R. C., Prausnitz, J. M., and Poling, B. F., *The Properties of Liquids and Gases,* 4th ed., McGraw-Hill, New York, 1987. With permission.

[b] Fetter, C. W., Jr., *Applied Hydrogeology,* Charles E. Merrill Publishing, Columbus, OH, 1980. With permission.

in site remediation projects. In general, the size of the plume in the groundwater would be much larger than that in the vadose zone because of the spread of the dissolved plume in the groundwater. If the water table fluctuates, the capillary fringe will move upward or downward with the water table. Consequently, the capillary fringe above the dissolved groundwater plume can become contaminated. In addition, if free-floating product exists, the fluctuation of the water table will cause the free product to move away vertically and laterally. The site remediation for this scenario will be more complicated and difficult.

The height of capillary fringe at a site strongly depends on its subsurface geology. For pure water at 20°C in a clean glass tube, the height of capillary rise can be approximated by the following equation:

$$h_c = \frac{0.153}{r} \qquad \text{[Eq. II.1.9]}$$

where h_c is the height of capillary rise in centimeters, and r is the radius of the capillary tube in centimeters. This formula can be used to estimate the height of the capillary fringe. As shown in Eq. II.1.9, the thickness of the capillary fringe will vary inversely with the pore size of a formation. Table II.1.B summarizes the information from two references with regard to capillary fringe. As the grain size becomes smaller, the pore radius gets smaller, and capillary rise increases. The capillary fringe of a clayey aquifer can exceed 10 ft.

Example II.1.5 *Thickness of the capillary fringe*

A core sample was taken from a contaminated unconfined aquifer and analyzed for pore size distributions. The effective pore size was determined to be 5 μm. Estimate the thickness of the capillary fringe of this aquifer.

Solution:

Pore size = 5×10^{-6} m = 5×10^{-4} cm.

Using Eq. II.1.9, we obtain capillary rise = $(0.153)/(5 \times 10^{-4})$ = 306 cm
= 3.06 m = 10.0 ft

Discussion. The units of h_c and r in Eq. II.1.9 are in centimeters.

II.1.6 Estimating the mass and volume of the free-floating product

The product leaked from a UST may accumulate on top of the water table and form a nonaqueous phase liquid (NAPL) layer. For site remediation it is often necessary to estimate the volume or mass of this free-floating product. The thickness of free product found in the monitoring wells has been directly used to calculate the volume of the free product. However, these calculated values are seldom representative of the actual free product volume existing in the formation.

It is now well known that the thickness of free product found in the formation (the actual thickness) is much smaller than that floating on top of the water in a monitoring well (the apparent thickness). Using the apparent thickness, without any adjustment, to estimate the volume of free product may lead to an overestimate of the free product volume and overdesign of the remediation system. The overestimate of free product in the RI phase may cause difficulties in obtaining approval for final site closure because the remedial action can never recover the full amount of free product reported in the site assessment report.

Factors affecting the difference between the actual thickness and the apparent thickness include the densities (or specific gravity) of the free product and the groundwater and the characteristics of the formation (especially the pore sizes). Several approaches have been presented in the literature to correlate these two thickness. Recently, Ballestero et al.[1] developed an equation using heterogeneous fluid flow mechanics and hydrostatics to determine the actual free product thickness in an unconfined aquifer. The equation is

$$t_g = t(1 - S_g) - h_a \qquad \text{[Eq. II.1.10]}$$

where t_g = actual (formation) free product thickness, t = apparent (wellbore) product thickness, S_g = specific gravity of free product, and h_a = distance from the bottom of the free product to the water table.

If no further data for h_a are available, average wetting capillary rise can be used as h_a. Information on capillary rise can be found in Section II.1.5.

To estimate the actual thickness of free product, the following procedure can be used:

Step 1: Determine the specific gravity of free product. (The specific gravity of gasoline can be reasonably assumed as 0.75 to 0.85 if no additional information is available.)
Step 2: Determine the apparent thickness of the free product in the well.
Step 3: Determine the actual thickness of free product in the formation by inserting values of the above parameters into Eq. II.1.10.

Information needed for this calculation

- Specific gravity (or density) of the free product, S_g
- Measured thickness of free product in the well, t
- Capillary rise, h_c

To determine the mass and volume of the free-floating product the following procedure can be used:

Step 1: Determine the areal extent of the free-floating product.
Step 2: Determine the true thickness of the free-floating product.
Step 3: Determine the volume of the free-floating product by multiplying the area with the true thickness and the porosity of the formation.
Step 4: Determine the mass of the free-floating product by multiplying the volume with its density.

Information needed for this calculation

- Areal extent of the free-floating product
- True thickness of the free-floating product
- Porosity of the formation
- Density (specific gravity) of the free-floating product

Example II.1.6A *Determine the true thickness of the free-floating product*

A recent survey of a groundwater monitoring well showed a 75-in thick layer of gasoline floating on top of the water. The density of gasoline is 0.8 g/cm^3, and the thickness of the capillary fringe above the water table is 1 ft. Estimate the actual thickness of the free-floating product in the formation.

Solution:

Using Eq. II.1.10, we obtain:

Actual free product thickness = (75)(1 – 0.8) – 12 = 3 in

Discussion. As shown in this example, the actual thickness of the free product is only 3 in, while the apparent thickness within the monitoring well is much higher at 75 in.

Example II.1.6B *Estimate the mass and volume of the free-floating product*

Recent groundwater monitoring results at a contaminated site indicate the areal extent of the free-floating product is approximately a rectangular shape of 50 ft × 40 ft. The true thicknesses of the free-floating product in the four monitoring wells inside the plume are 2, 2.6, 2.8, and 3 ft, respectively. The porosity of the subsurface is 0.35. Estimate the mass and volume of the free-floating product present at the site. Assume the specific gravity of the free-floating product is 0.8.

Solution:

a. The areal extent of the free-floating product = (50′)(40′) = 2000 ft^2.

b. The average thickness of the free-floating product

$$= (2 + 2.6 + 2.8 + 3)/4 = 2.6 \text{ ft}$$

c. The volume of the free-floating product = (area)(thickness)(porosity of the formation)

$$= (2000 \text{ ft}^2)(2.6 \text{ ft})(0.35) = 1820 \text{ ft}^3$$

$$= (1820 \text{ ft}^3)(7.48 \text{ gal/ft}^3) = 13{,}610 \text{ gal}$$

d. Mass of the free-floating product = (volume of the free-floating product)(density of the free-floating product)

$$= (1820 \text{ ft}^3)\{0.8 \text{ g/cm}^3)[(62.4 \text{ lb/ft}^3)/(1 \text{ g/cm}^3)]\}$$

$$= 90{,}854 \text{ lb} = 41{,}300 \text{ kg}$$

II.1.7 *Determination of the extent of contamination — a comprehensive example calculation*

This subsection presents a comprehensive example related to the assessment of a contaminated site starting from tank pull, soil boring, and groundwater monitoring.

Example II.1.7 *Determination of the extent of contamination*

A gasoline station is located in the greater Los Angeles Basin within the floor plain of the Santa Ana River. The site is underlain primarily with coarser-grained river deposit alluvium. Three 5000-gal steel tanks were excavated and removed in May of 1997, with the intention that they would be replaced with three dual-wall fiberglass tanks within the same excavation.

During the tank removal it was observed that the tank backfill soil exhibited a strong gasoline odor. Based on visual observations, the fuel hydrocarbon in the soil appeared to have been caused by overspillage during filling at unsealed fill boxes or minor piping leakage at the eastern end of the tanks. The excavation resulted in a pit of 20′ × 30′ × 18′ (L × W × H). The excavated soil was stockpiled on site. Four samples were taken from the piles and analyzed for TPH using EPA method 8015. The TPH concentrations were ND (not detectable, <10), 200, 400, and 800 ppm, respectively.

The tank pit was then backfilled with clean dirt and compacted. Six vertical soil borings (two within the excavated area) were drilled to characterize the subsurface geological condition and to delineate the plume. The borings were drilled using the hollow-stem-auger method. Soil samples were taken by a 2″-diameter split-spoon sampler with brass soil sample retainers every 5 ft bgs. The water table is at 50 ft bgs, and all the borings were terminated at 70 ft bgs. All the borings were then converted to 4-in groundwater monitoring wells.

Selected soil samples from the borings were analyzed for TPH and BTEX (EPA method 8020). The analytical results indicated that the samples from the borings outside the excavated area were all ND. The other results are listed below:

Boring No.	Depth (ft)	TPH (ppm)	Benzene (ppb)	Toluene (ppb)
B1	25	800	10,000	12,000
B1	35	2000	25,000	35,000
B1	45	500	5,000	7500
B2	25	<10	<100	<100
B2	35	1200	10,000	12,000
B2	45	800	2000	3000

It was also found that free-floating gasoline product was present in the two monitoring wells located within the excavated area. The apparent thickness of the product in these two wells was converted to its actual thickness in the formation as 1 and 2 ft, respectively. The porosity and bulk density of both soil and aquifer matrices are 0.35 and 1.8 g/cm^3, respectively.

Assuming that the leakage contaminated a rectangular block defined by the bottom of the tank pit and the surface of the water table, with length and width equal to those of the tank pit, estimate the following:

a. Total volume of the soil stockpiles (in cubic yards)
b. Mass of TPH in the stockpiles (in kilograms)
c. Volume of the contaminated soil left in the vadose zone (in cubic meters)
d. Mass of TPH, benzene, and toluene in the vadose zone (in kilograms)
e. Mass fraction and mole fraction of benzene and toluene in the leaked gasoline
f. Volume of the free product (in gallons)
g. Total volume of gasoline leaked (in gallons) [Note: neglect the dissolved phase in the underlying aquifer]

Solution:

a. Total volume of the soil stockpiles

$$= [(\text{volume of tank pit}) - (\text{volume of USTs})](\text{soil fluffy factor})$$

$$= [(20' \times 30' \times 18') - (3)(5000\text{ gal})(\text{ft}^3/7.48\text{ gal})](1.15)$$

$$= (8795\text{ ft}^3)(1.15) = 10{,}100\text{ ft}^3 = 375\text{ yd}^3$$

b. Mass of TPH in the stockpiles = $(V)(\rho_b)(C) = (M_s)(C)$.

$$C = (10 + 200 + 400 + 800)/4 = 352.5\text{ mg/kg}$$

$$\rho_b = (1.8\text{ g/cm}^3)\ (28{,}317\text{ cm}^3/\text{ft}^3)(\text{kg}/1000\text{ g})$$

$$= 51.0\text{ kg/ft}^3 = 1376\text{ kg/yd}^3$$

Mass of TPH in the stock piles

$$= (8795\text{ ft}^3)(51.0\text{ kg/ft}^3)\ (352.5\text{ mg/kg})(\text{kg}/10^6\text{ mg}) = 158\text{ kg}$$

c. Volume of contaminated soil left in the vadose zone

$$= (20' \times 30')(50' - 18') = 19{,}200\text{ ft}^3$$

$$= (19{,}200\text{ ft}^3)(0.0283\text{ m}^3/\text{ft}^3) = 544\text{ m}^3$$

d. Mass of TPH, benzene, and toluene in the vadose zone

$$= (V)(\rho_b)(C) = (M)(C) \text{ or using a more precise approach}$$

$$\sum_i (A_i)(h_i)(\rho_b)(C_i)$$

	Average concentration (mg/kg)	Mass (kg)
TPH	(800 + 2000 + 500 + 10 + 1200 + 800)/ 6 = 885	(19,200)(51)(885)/1,000,000 = 866
Benzene	(10 + 25 + 5 + 0.1 + 10 + 2)/6 = 8.68	(19,200)(51)(8.68)/1,000,000 = 8.50
Toluene	(12 + 35 + 7.5 + 0.1 + 12 + 3)/6 = 11.6	(19,200)(51)(11.6)/1,000,000 = 11.34

e. Mass fraction and mole fraction of benzene and toluene:

	Mass (kg)	Mass fraction	Mol. wt.	kg-mole	Mole fraction
TPH	866		100	866/100 = 8.66	
Benzene	8.50	8.50/866 = 0.0098	78	8.50/78 = 0.109	0.109/8.66 = 0.0126
Toluene	11.34	11.3/866 = 0.0130	92	11.3/92 = 0.123	0.123/8.66 = 0.0142

f. Volume of the free-floating product

$$= (h)(A)(\phi) = [(1 + 2)/2](20 \times 30)(0.35)$$

$$= 315 \text{ ft}^3 \times (7.48 \text{ gal/ft}^3) = 2360 \text{ gal}$$

Mass of free-floating product

$$= (V)(\rho) = (2360 \text{ gal})(3.785 \text{ L/gal})(0.75 \text{ kg/L}) = 6700 \text{ kg}$$

g. Total volume of gasoline leaked = Sum of those in excavated soil, vadose zone, free product, and dissolved phase

$$= 158 + 866 + 6700 = 7724 \text{ kg}$$
(neglecting the dissolved phase)

$$= 7724 \text{ kg}/(0.75 \text{ kg/L})$$

$$= 10{,}300 \text{ L} = (10{,}300/3.785) \text{ gal} = 2720 \text{ gal}$$

Discussion. Determination of contaminant mass in the aquifer will be covered in Section II.3.

II.2 Soil borings and groundwater monitoring wells

This section deals with calculations related to installation of soil borings and groundwater monitoring wells and purging before groundwater sampling.

II.2.1 Amount of cuttings from soil boring

The cuttings from soil borings are often temporarily stored on site in 55-gal drums before final disposal. It becomes necessary to estimate the amount of cuttings and the number of drums needed. The calculation is relatively straightforward and easy, as shown below.

To estimate the amount of cuttings from soil boring, the following procedure can be used:

Step 1: Determine the diameter of the boring, $d_{b.}$
Step 2: Determine the depth of the boring, h.
Step 3: Calculate the volume of the cutting using the following formula:

$$\text{Volume of cuttings} = \sum \left(\frac{\pi}{4} d_b^2\right)(h)(\text{fluffy factor}) \quad \text{[Eq. II.2.1]}$$

Information needed for this calculation

- Diameter of each boring, d_b
- Depth of the each boring, h
- Soil fluffy factor

Example II.2.1 Amount of cuttings from soil boring

Four 10-in boreholes are drilled to 50 ft below ground surface level for installation of 4-in groundwater monitoring wells. Estimate the amount of soil cuttings and determine the number of 55-gal drums needed to store the cuttings.

Solution:

a. Volume of cuttings from each boring

$$= [(\pi/4)(10/12)^2](50)(1.1) = 30.0 \text{ ft}^3$$

Volume of cutting from all borings

$$= (4)(30.0) = 120 \text{ ft}^3$$

b. Number of 55-gal drums needed

$$= (120.0 \text{ ft}^3)(7.48 \text{ gal/ft}^3) \div (55 \text{ gallon/drum}) = 16.3 \text{ drums}$$

Answer: Seventeen 55-gal drums needed.

II.2.2 Amount of packing materials and/or bentonite seal

Packing and seal materials need to be purchased and shipped to the site before installation of monitoring wells. A good estimate of the amount of packing material and bentonite seal is necessary for site remediation.

To estimate the packing and seal materials needed, the following procedure can be used:

Step 1: Determine the diameter of the boring, d_b.
Step 2: Determine the diameter of the well casing, d_c.
Step 3: Determine the depth of the well packing or bentonite seal, h.
Step 4: Calculate the volume of the packing or bentonite seal using the following formula:

$$\text{Volume of packing or bentonite needed} = \frac{\pi}{4}(d_b^2 - d_c^2)h \qquad \text{[Eq. II.2.2]}$$

Step 5: Determine the mass of the well packing or bentonite needed by multiplying its volume by its bulk density.

Information needed for this calculation

- Diameter of the borehole, d_b
- Diameter of the casing, d_c
- Depth of the packing or bentonite seal, h
- Bulk density of the packing or bentonite seal, ρ_b

Example II.2.2A Amount of packing materials need

The four monitoring wells in Example II.2.1 are installed 15 ft into the groundwater aquifer. The wells are perforated (0.02-in slot opening) 15 ft below and 10 ft above the water table. Monterey Sand #3 is selected as the packing material. Estimate the number of 50-lb sand bags needed for this application. Assume the bulk density of sand to be 1.8 g/cm^3 (112 lb/ft^3).

Solution:

a. Packing interval for each well

$$= \text{perforation interval} + 1 \text{ ft} = (10 + 15) + 1 = 26 \text{ ft}$$

Volume of sands needed for each well

$$= \{(\pi/4)[(10/12)^2 - (4/12)^2]\}(26) = 11.9 \text{ ft}^3$$

Volume of sands needed for four wells

$$= (4)(11.9) = 47.6 \text{ ft}^3$$

b. Number of 50-lb sand bags needed = $(47.6 \text{ ft}^3)(112 \text{ lb/ft}^3) \div (50 \text{ lb/bag})$ = 107 bags

Answer: 107 bags needed.

Discussion

1. Packing interval should be slightly larger than the perforation interval.
2. The outside diameter of 4-in well casing should be slightly larger than 4 in. Theoretically, it will make the calculated volume of sand be slightly larger than the actual volume.
3. We should add an additional 10% to the estimate of sand usage as a safety factor to take into consideration that bore hole shape would not be a perfect cylinder.

Example II.2.2B Amount of bentonite seal need

The four monitoring wells in Example II.2.2 are sealed with 5 ft of bentonite below the top grout. Estimate the number of 50-lb bags of bentonite needed for this application. Assume the bulk density of bentonite to be 1.8 g/cm^3 (112 lb/ft^3).

Solution:

a. Volume of bentonite needed for each well

$$= \{(\pi/4)[(10/12)^2 - (4/12)^2]\}(5) = 2.29 \text{ ft}^3$$

Volume of bentonite needed for four wells

$$= (2.29)(4) = 9.16 \text{ ft}^3$$

b. Number of 50-lb bentonite bags needed = $(9.16 \text{ ft}^3)(112 \text{ lb/ft}^3) \div (50 \text{ lb/bag})$ = 20.5 bags.

Answer: 21 bags are needed.

Discussion

1. The outside diameter of 4-in well casing should be slightly larger than 4 in. Theoretically, it will make the calculated volume of bentonite be slightly larger than the actual volume.
2. We should add an additional 10% in the estimate of bentonite usage as a safety factor to take into consideration that bore hole shape would not be a perfect cylinder.

II.2.3 Well volume for groundwater sampling

Purging is the process of removing stagnant water from a monitoring well before sampling groundwater. The stagnant volume includes the water inside the well casing and in the sandpack. A few parameters are often monitored, such as conductivity, pH, and temperature, to ensure they reach a consistent end point before sampling. There is no universally correct purge volume. The purge volume is site specific and depends heavily on the subsurface geology. A rule of thumb of purging three to five well volumes before groundwater sampling can be a starting point. The purged water is often contaminated and needs to be treated, stored, disposed of off site. A good estimate of the volume of purged water is necessary for site remediation.

To estimate the amount of purged water the following procedure can be used:

Step 1: Determine the diameter of the boring, $d_{b.}$
Step 2: Determine the diameter of the well casing, $d_{c.}$
Step 3: Determine the depth of the water in the well, h.
Step 4: Calculate the well volume using the following formula:

Well volume = volume of the groundwater enclosed inside the well casing + volume of the groundwater in the pore space of the packing

$$\text{Well volume} = \left[\frac{\pi}{4}d_c^2\right]h + \left[\frac{\pi}{4}(d_b^2 - d_c^2)h\right]\phi \qquad \text{[Eq. II.2.3]}$$

Information needed for this calculation

- Diameter of the borehole, d_b
- Diameter of the casing, d_c
- Porosity of the packing, ϕ
- Depth of the well water, h

Example II.2.3 Well volume for groundwater sampling

The water depth inside one of the four monitoring wells in Example II.2.2 was measured to be 14.5 ft. Three well volumes need to be purged out before sampling. Calculate the amount of purge water and also the number of 55-gal drums needed to store the water. Assume the porosity of the well packing to be 0.40.

Solution:

a. Well volume

$$= (\pi/4)(4/12)^2(14.5) + (\pi/4)[(10/12)^2 - (4/12)^2](14.5)(0.4) = 3.92 \text{ ft}^3$$

b. Three well volumes = (3)(3.92) = 11.8 ft^3 = 88 gal.

c. Number of 55-gal drums needed

$$= (11.8 \text{ ft}^3)(7.48 \text{ gal/ft}^3) \div (55 \text{ gal/drum}) = 1.6 \text{ drums.}$$

Answer: Two 55-gal drums are needed.

II.3 *Mass of contaminants present in different phases*

Once an NAPL enters a vadose zone, it may end up in four different phases. Molecules may leave the free product and enter the air void. The compound in the air and/or in the free product, in contact with the soil moisture, may dissolve in the liquid. The compound in the air, in the free product, and in the soil moisture may adsorb onto the soil grains. In other words, the NAPL can partition into four phases: (1) free product, (2) vapor in the void, (3) dissolved constituent in soil moisture, and (4) adsorbed onto the soil grains. The concentrations of the contaminant in the air void, in the soil moisture, and on the soil grains are interrelated and affected greatly by the presence or absence of the free product. The partition of the contaminants in these four phases has a great impact on the fate and transport of the compound and the required site remediation effort. Good understanding of this partition phenomenon is necessary to implement cost-effective alternatives for the site cleanup.

In this section, we will first discuss the vapor concentration resulting from the presence of free-product in the pores (Section II.3.1). We will then describe the relationship between the contaminant concentration in the liquid and that in the air (Section II.3.2). The relationship between the contaminant concentration in the liquid and that in the soil will be covered next (Section II.3.3). The relationship among the liquid, vapor, and solid concentrations will then be discussed (Section II.3.4). The last subsection describes the procedure to determine the partition of contaminant in these phases (Section II.3.5).

II.3.1 Equilibrium between free product and vapor

When a liquid is in contact with air, molecules in the liquid will tend to enter the air phase as a vapor, via evaporation or volatilization. The vapor pressure of a liquid is the pressure exerted by its vapor at equilibrium. It is usually measured in millimeters of mercury (760 mmHg = 760 torr = 1 atm = 1.013 × 10^5 N/m^2 = 1.013 × 10^5 Pascal = 14.696 psi) and varies greatly with temperature. In general, the higher the temperature, the higher the vapor pressure. Several equations have been established to correlate the vapor pressure and temperature; the Clausius–Clapeyron equation is commonly used. This equation assumes that the enthalpy of vaporization is independent of temperature, and is expressed as

$$\ln \frac{P_1^{sat}}{P_2^{sat}} = -\frac{\Delta H^{vap}}{R}\left[\frac{1}{T_1} - \frac{1}{T_2}\right] \qquad \text{[Eq. II.3.1]}$$

where P^{sat} is the vapor pressure of the compound as a pure liquid, T is the absolute temperature, R is the universal gas constant, and ΔH^{vap} is the enthalpy of vaporization, which can be found in chemistry handbooks (see Reference 9). Table II.3.A lists the values of the universal gas constant in various units.

The Antoine equation is an empirical equation widely used and has the following form:

$$\ln P^{sat} = A - \frac{B}{T + C} \qquad \text{[Eq. II.3.2]}$$

where A, B, and C are the Antoine constants, which can be found in chemistry handbooks (see Reference 10).

For an ideal liquid mixture, the vapor–liquid equilibrium follows Raoult's law as

$$P_A = (P^{vap})(x_A) \qquad \text{[Eq. II.3.3]}$$

where P_A = partial pressure of compound A in the vapor phase, P^{vap} = vapor pressure of compound A as a pure liquid, and x_A = mole fraction of compound A in the liquid phase.

Table II.3.A Values of the Universal Gas Constants

R = 82.05 (cm^3 · atm)/(g mol)(K)	= 83.14 (cm^3 · bar)/(g mol)(K)
= 8.314 (J)/(g mol)(K)	= 1.987 (cal)/(g mol)(K)
= 0.7302 (ft^3 · atm)/(lb mol)(R)	= 10.73 (ft^3 · psia)/(lb mol)(R)
= 1545 (ft · lb$_f$)/(lb mol)(R)	= 1.987 (Btu)/(lb mol)(R)

The partial pressure is the pressure that a compound would exert if all other gases were not present. This is equivalent to the mole fraction of the compound in the gas phase multiplied by the entire pressure of the gas. Raoult's law holds only for ideal solutions. In dilute aqueous solutions commonly found in environmental applications, Henry's law, which will be discussed in the next section, is more suitable.

Example II.3.1A *Vapor concentration in void with presence of free product*

Benzene leaked from a UST at a site and entered the vadose zone. Estimate the maximum benzene concentration (in ppmV) in the pore space of the subsurface. The temperature of the subsurface is 25°C.

Solution:

From Table II.1.A, the vapor pressure of benzene is 95 mmHg at 25°C.
95 mmHg = (95 mmHg) ÷ (760 mmHg/1 atm) = 0.125 atm.
The partial pressure of benzene in the pore space is 0.125 atm (125,000 × 10^{-6} atm), which is equivalent to 125,000 ppmV.

Discussion. The 125,000 ppmV is the vapor concentration in equilibrium with the pure benzene liquid. The equilibrium can occur in a confined space or a stagnant phase. If the medium is not totally confined, the vapor tends to move away from the source and creates a concentration gradient (the vapor concentration decreases with the distance from the free liquid). However, in the vicinity of the free product, the vapor concentration would be at or near this equilibrium value.

Example II.3.1B *Using the Clausius–Clapeyron equation to estimate the vapor pressure*

The enthalpy of vaporization of benzene is 33.83 kJ/mol,[9] and the vapor pressure of benzene at 25°C is 95 mmHg (from Table II.1.A). Estimate the vapor pressure of benzene at 20°C using the Clausius–Clapeyron equation.

Solution:

Heat of vaporization = 33.83 kJ/mol = 33,830 J/mol. R = 8.314 (J)/(g mol)(K) from Table II.3.A. Using Eq. II.3.1, we obtain

$$\ln\frac{95}{P_2^{sat}} = -\frac{33,830}{8.314}\left[\frac{1}{(273+25)} - \frac{1}{(273+20)}\right]$$

Answer: P^{sat} of benzene at 20°C = 75 mmHg

Discussion. As expected, the vapor pressure of benzene at 20°C is lower than that at 25°C. The difference is approximately 20%.

Example II.3.1C *Using the Antoine equation to estimate the vapor pressure*

The empirical constants of the Antoine equation for benzene are[10] A = 15.9008, B = 2788.51, and C = –52.36. Estimate the vapor pressure of benzene at 20 and at 25°C using the Antoine equation.

Solution:

a. Use Eq. II.3.2, at 20°C

$$\ln P^{sat} = A - \frac{B}{T+C} = 15.9008 - \frac{2788.51}{(293.18 - 52.36)} = 4.322$$

$$\text{So, } P^{vap} = 75.3 \text{ mmHg}$$

b. Use Eq. II.3.2, at 25°C

$$\ln P^{sat} = A - \frac{B}{T+C} = 15.9008 - \frac{2788.51}{(298.18 - 52.36)} = 4.557$$

$$\text{So, } P^{vap} = 95.3 \text{ mmHg}$$

Discussion

1. The calculated benzene vapor pressure, 95.3 mmHg (at 25°C) is essentially the same as that in Table II.1.A, 95 mmHg.
2. The calculated benzene vapor pressure, 75.3 mmHg (at 20°C) is essentially the same as that in Example II.3.1B, 75 mmHg, which uses the Clausius–Clapeyron equation.

Example II.3.1D *Vapor concentration in void with presence of free product*

An industrial solvent, consisting of 50% (by wt) toluene and 50% ethylbenzene, leaked from a UST at a site and entered the vadose zone. Estimate the maximum toluene and ethylbenzene concentrations (in ppmV) in the pore space of the subsurface. The temperature of the subsurface is 20°C.

Solution:

From Table II.1.A, the vapor pressure of toluene (C_7H_8, molecular weight = 92) is 22 mmHg and that of ethylbenzene (C_8H_{10}, molecular weight = 106) is 7 mmHg at 20°C. For 50% by weight of toluene, the corresponding percentage by moles will be

(moles of toluene) ÷ [(moles of toluene) + (moles of ethylbenzene)] × 100

$$= (50/92) \div [(50/92) + (50/106)] \times 100 = 53.5\%$$

The partial pressure of toluene in the pore space can be determined from Eq. II.3.3:

$$(22)(0.535) = 11.78 \text{ mmHg} = 0.0155 \text{ atm} = 15{,}500 \text{ ppmV}$$

The partial pressure of ethylbenzene in the pore space can be determined from Eq. II.3.3:

$$(7)[1 - (0.535)] = 3.25 \text{ mmHg} = 0.0043 \text{ atm} = 4{,}300 \text{ ppmV}$$

Discussion. The vapor concentrations are those in equilibrium with the pure solvent. The equilibrium can occur in a confined space or a stagnant phase. If the medium is not totally confined, the vapor tends to move away from the source and creates a concentration gradient (the vapor concentration decreases with the distance from the free liquid). However, in the vicinity of the free product, the vapor concentration would be at or near the equilibrium value.

II.3.2 Liquid–vapor equilibrium

The contaminant in the void of the vadose zone will tend to enter the liquid, via dissolution or absorption. Equilibrium conditions exist when the rate of contaminant entering the liquid equals the rate of contaminant volatilizing from the liquid.

Henry's law is used to describe the equilibrium relationship between the liquid concentration and vapor concentration. At equilibrium, the partial pressure of a gas above a liquid is proportional to the concentration of the chemical in the liquid. Henry's law can be expressed as

$$P_A = H_A C_A \qquad \text{[Eq. II.3.4]}$$

where P_A = partial pressure of compound A in the gas phase, H_A = Henry's constant of compound A, and C_A = concentration of compound A in the liquid phase.

It should be noted that in some of the air pollution books or references, Henry's law is written as $C_A = H_A P_A$. This Henry's constant is the inverse of the one used in this book and most of the site remediation applications.

Henry's law can also be expressed in the following form:

$$G = HC \qquad \text{[Eq. II.3.5]}$$

where C is the contaminant concentration in the liquid phase and G is the concentration in the gas phase.

Henry's law has been widely used in various disciplines to describe the distribution of solute in vapor phase and liquid phase. The units of the Henry's law constant (or Henry's constant) reported in the literature vary considerably. The units commonly encountered are atm/mole fraction, atm/M, M/atm, atm/(mg/L), and dimensionless. When inserting the value of Henry's constant into the two equations above, it is important to check if its dimensions match the dimensions of the other two parameters. Process designers with whom I have conferred normally use the units they are familiar with and often have difficulty performing the necessary unit conversions. For your convenience, Table II.3.B is the conversion table for Henry's constant. Use of Henry's constant in dimensionless form has increased significantly. It should be noted that it is not a "(mole fraction)/(mole fraction)" dimensionless unit. The actual meaning of the Henry's constant in dimensionless form is (concentration of solute in vapor phase)/(concentration in liquid phase), which can be either (M/M) or [(mg/L)/(mg/L)]. To be more precise, it has a unit of "(unit volume of liquid)/(unit volume of vapor)."

Henry's constant of any given compound varies with temperature. The Henry's constant is practically the ratio of the vapor pressure divided by solubility, provided that both are measured at the same temperature, that is,

$$H = \frac{\text{vapor pressure}}{\text{solubility}} \qquad \text{[Eq. II.3.6]}$$

This equation implies that the higher the vapor pressure the larger the Henry's constant is. In addition, the lower the solubility (or less soluble compound), the larger Henry's constant will be. For most organic compounds, the vapor pressure increases and the solubility decreases with tem-

Table II.3.B Henry's Constant Conversion Table

Desired unit for Henry's constant	Conversion equation
atm/M, or atm L/mole	$H = H^*RT$
atm m^3/mole	$H = H^*RT/1000$
M/atm	$H = 1/(H^*RT)$
atm/(mole fraction in liquid), or atm	$H = (H^*RT)[1000\gamma/W]$
(mole fraction in vapor)/(mole fraction in liquid)	$H = (H^*RT)[1000\gamma/W]/P$

Note: H^* = Henry's constant in dimensionless form, γ = specific gravity of the solution (1 for dilute solution), W = equivalent molecular weight of solution (18 for dilute aqueous solution), R = 0.082 atm/(K)(M), T = system temperature in Kelvin, P = system pressure in atm (usually = 1 atm), and M = solution molarity in (g mol/L).

From Kuo, J. F. and Cordery, S. A., Discussion of nomograph for air stripping of VOC from water, *J. Environ. Eng.*, V. 114, No. 5, p. 1248–1250, 1988. With permission.

perature. Consequently, Henry's constant, as defined by Eq. II.3.4 or Eq. II.3.5, should increase with temperature.

Table II.3.C summarizes the values of Henry's constant, vapor pressure, and solubility of commonly found contaminants. The values for the octanol–water partition coefficient, K_{ow}, and the diffusion coefficients, D, are also listed, and discussions on these parameters will be given in later sections. For more complete lists of these values, chemistry handbooks and references should be consulted.

Example II.3.2A *Unit conversions for Henry's constant*

As shown in Table II.3.C, the Henry's constant for benzene in water at 25°C is 5.55 atm/M. Convert this value to dimensionless units and also to units of atm.

Solution:

From Table II.3.B

$$H = H^*RT = 5.55 = H^*\ (0.082)(273 + 25)$$
$$H^* = 0.227\ \text{(dimensionless)}$$

Also, from Table II.3.B

$$H = (H^*RT)[1000\gamma/W]$$
$$H = [(0.227)(0.082)(273 + 25)][(1000)(1)/(18)] = 308.3\ \text{atm}$$

Discussion. As mentioned earlier, the dimensionless Henry's constant is becoming more popular. Benzene is a VOC of concern and is shown in most, if not all, databases of Henry's constant values. It may not be a bad idea to memorize that benzene has a dimensionless Henry's constant of 0.23 at ambient conditions. To convert the Henry's constant of another compound in the database, just multiply the ratio of the Henry's constant (in any units) of that compound and of benzene by 0.23. For example, to find the dimensionless Henry's constant of methylene chloride, first read the Henry's constant for methylene chloride, 2.03, and for benzene, 5.55, from Table II.3.C. Then find the ratio of these two and multiply by 0.23 as $[(2.03)/(5.55)] \times (0.23) = 0.084$.

Example II.3.2B *Estimate Henry's constant from solubility and vapor pressure*

As shown in Table II.3.C, the vapor pressure of benzene is 95.2 mmHg and its solubility in water is 1780 mg/L at 25°C. Estimate the Henry's constant from the given information.

Table II.3.C Physical Properties of Common Contaminants

Compound	MW (g/mole)	*H* (atm/M)	p_{vap} (mmHg)	*D* (cm^2/s)	Log K_{ow}	Solubility (mg/L)	T (°C)
Benzene	78.1	5.55	95.2	0.092	2.13	1780	25
Bromomethane	94.9	106		0.108	1.10	900	20
2-Butanone	72	0.0274			0.26	268,000	
Carbon disulfide	76.1	12	260		2.0	2940	20
Chlorobenzene	112.6	3.72	11.7	0.076	2.84	488	25
Chloroethane	64.5	14.8			1.54	5740	25
Chloroform	119.4	3.39	160	0.094	1.97	8000	20
Chloromethane	50.5	44			0.95	6450	20
Dibromochloromethane	208.3	2.08			2.09	0.2	
Dibromomethane	173.8	0.998				11,000	
1,1-Dichloroethane	99.0	4.26	180	0.096	1.80	5500	20
1,2-Dichloroethane	99.0	0.98	61		1.53	8690	20
1,1-Dichloroethylene	96.9	34	600	0.084	1.84	210	25
1,2-Dichloroethylene	96.9	6.6	208		0.48	600	20
1,2 Dicholopropane	113.0	2.31	42		2.00	2700	20
1,3-Dichloropropylene	111.0	3.55	38		1.98	2800	25
Ethylbenzene	106.2	6.44	7	0.071	3.15	152	20
Methylene chloride	84.9	2.03	349		1.3	16,700	25
Pyrene	202.3	0.005			4.88	0.16	26
Styrene	104.1	9.7	5.12	0.075	2.95	300	20

Table II.3.C ***(continued)*** Physical Properties of Common Contaminants

Compound	MW (g/mole)	H (atm/M)	P_{vap} (mmHg)	D (cm^2/s)	Log K_{ow}	Solubility (mg/L)	T (°C)
1,1,1,2-Tetrachloroethane	167.8	0.381	5	0.077	3.04	200	20
1,1,2,2-Tetrachloroethane	167.8	0.38			2.39	2900	20
Tetrachloroethylene	165.8	25.9		0.077	2.6	150	20
Tetrachloromethane	153.8	23			2.64	785	20
Toluene	92.1	6.7	22	0.083	2.73	515	20
Tribromoethane	252.8	0.552	5.6		2.4	3200	30
1,1,1-Trichloroethane	133.4	14.4	100		2.49	4400	20
1,1,2-Trichloroethane	133.4	1.17	32		2.47	4500	20
Trichloroethylene	131.4	9.1	60		2.38	1100	25
Trichlorofluoromethane	137.4	58	667	0.083	2.53	1100	25
Vinyl chloride	62.5	81.9	2660	0.114	1.38	1.1	25
Xylenes	106.2	5.1	10	0.076	3.0	198	20

From U.S. EPA, *CERCLA Site Discharges to POTWs Treatability Manual*, EPA 540/2-90-007, Office of Water, U.S. EPA, Washington, DC, 1990.

From LaGrega, M.D. , Buckingham, P.L., and Evans, J.C., *Hazardous Waste Management*, McGraw-Hill, New York, 1994. With permission.

Solution:

From Eq. II.3.6, we know that Henry's constant is the ratio between the vapor pressure and the solubility, so

$$H = (95.2 \text{ mmHg})/(1780 \text{ mg/L}) = 0.0535 \text{ mmHg}/(\text{mg/L})$$

To compare with the value given in Table II.3.C, we need to convert the units of the vapor pressure and solubility.

$$P^{vap} = 95.2/760 = 0.125 \text{ atm}$$

$$S = 1780 \text{ mg/L} = 1.78 \text{ g/L} = (1.78 \text{ g/L}) \div (78.1 \text{ g/g mol})$$
$$= 0.0228 \text{ mol/L} = 0.0228 \text{ M}$$

So

$$H = (0.125 \text{ atm}) \div (0.0228 \text{ M}) = 5.48 \text{ atm/M}$$

Discussion. The calculated value, 5.48, is essentially the same as the value in Table II.3.C.

Example II.3.2C *Use Henry's law to calculate the equilibrium concentrations*

The subsurface of a site is contaminated with tetrachloroethylene (PCE). A recent soil vapor survey indicates that the soil vapor contained 1250 ppmV of PCE. Estimate the PCE concentration in the soil moisture. Assume the subsurface temperature to be 20°C.

Solution:

a. From Table II.3.C, for PCE

$$H = 25.9 \text{ atm/M and MW} = 165.8$$

$$\text{Also, } 1250 \text{ ppmV} = 1{,}250 \times 10^{-6} \text{ atm} = P_A$$

Use Eq. II.3.4,

$$P_A = H_A C_A = 1.25 \times 10^{-3} \text{ atm} = (25.9 \text{ atm/M})C_A$$

So,

$$C_A = 4.82 \times 10^{-5} \text{ M} = (4.82 \times 10^{-5} \text{ mole/L})(165.8 \text{ g/mole})$$
$$= 8 \times 10^{-3} \text{ g/L} = 8 \text{ mg/L} = 8 \text{ ppm}$$

b. We can also use the dimensionless Henry's constant to solve this problem

$$H = H^*RT = 25.9 = H^* (0.082)(273 + 20)$$

$$H^* = 1.08 \text{ (dimensionless)}$$

Use Eq. II.1.1 to convert ppmV to mg/m^3

$$1250 \text{ ppmV} = (1250)[(165.8/24.05)] \text{ mg/m}^3$$
$$= 8620 \text{ mg/m}^3 = 8.62 \text{ mg/L}$$

Use Eq. II.3.5

$$G = HC = 8.62 \text{ mg/L} = (1.08)C$$

So,

$$C = 8 \text{ mg/L} = 8 \text{ ppm}$$

Discussion

1. The two approaches yield identical results.
2. Henry's constant of PCE is relatively high (five times higher than that of benzene). A concentration of 8 mg/L of PCE in water is in equilibrium with a relatively high vapor concentration of 1250 ppmV.

II.3.3 Solid–liquid equilibrium

Adsorption

Adsorption is the process in which a component moves from liquid phase to solid phase across the interfacial boundary. Adsorption is caused by interactions among three distinct components:

- Adsorbent (e.g., vadose zone soil, aquifer matrix, and activated carbon)
- Adsorbate (e.g., the contaminant)
- Solvent (e.g., soil moisture and groundwater)

In adsorption, the adsorbate is removed from the solvent and taken by the adsorbent. Adsorption is an important mechanism affecting the contaminant's fate and transport in an environmental medium.

Adsorption isotherms

For a system where solid phase and liquid phase coexist, an adsorption isotherm describes the equilibrium relationship between the liquid and solid phases. The "isotherm" indicates that the relationship is for a constant temperature.

The most popular isotherms are the Langmuir isotherm and the Freundlich isotherm. Both were derived in the early 1900s. The Langmuir isotherm has a theoretical basis, while the Freundlich is a semiempirical relationship. For a Langmuir isotherm, the concentration in the soil increases with increasing concentration in the groundwater until a maximum concentration in the soil is reached. The Langmuir isotherm can be expressed as follows:

$$X = X_{max} \frac{KC}{1+KC} \quad \text{[Eq. II.3.7]}$$

where X is the sorbed concentration, C is the liquid concentration, K is the equilibrium constant, and X_{max} is the maximum adsorbed concentration.

On the other hand, the Freundlich isotherm can be expressed as the following form:

$$X = KC^{1/n} \quad \text{[Eq. II.3.8]}$$

Both K and $1/n$ are empirical constants. These constants are different for different compounds. For a given compound, the values will also be different for different temperatures. When using the isotherms, one should ensure that the units among the parameters and the empirical constants are consistent.

Both isotherms are nonlinear. Incorporating the nonlinear Langmuir isotherm or Freundlich isotherm into the mass balance equation to simulate the contaminant's fate and transport will make the computer simulation more difficult or more time consuming. Fortunately, it was found that, in many environmental applications, the linear form of the Freundlich isotherm applies. It is called the linear adsorption isotherm, since $1/n = 1$, thus

$$X = KC \quad \text{[Eq. II.3.9]}$$

which simplifies the mass balance equation in a fate and transport model.

Partition coefficient

For soil–water systems, the linear adsorption isotherm is often written in the following form:

$$X = K_p C, \text{ thus} \qquad K_p = X/C \quad \text{[Eq. II.3.10]}$$

where K_p is called the partition coefficient that measures the tendency of a chemical to be adsorbed by soil or sediment from a liquid phase and describes how the chemical compound distributes (partitions) itself between the two media. Henry's constant, which was discussed earlier, can be viewed as the vapor–liquid partition coefficient.

For a given organic chemical compound, the partition coefficient is not the same for every soil. The dominant mechanism of organic adsorption is the hydrophobic bonding between the compound and the natural organics associated with the soil. It was found that K_p increases as the fraction of organic carbon, f_{oc}, increases in soil, thus

$$K_p = f_{oc} K_{oc} \quad \text{[Eq. II.3.11]}$$

The organic carbon partition coefficient, K_{oc}, can be considered as the partition coefficient for the organic compound into a hypothetical pure organic carbon phase. For soil that is not 100% organics, the partition coefficient is discounted by the factor, f_{oc}. Clayey soil is often associated with more natural organic matter and, thus, has a stronger adsorption potential for organic contaminants.

K_{oc} is actually a theoretical parameter, and it is the slope of experimentally determined K_p vs. f_{oc} curves. K_{oc} values for many compounds are not available. Much research has been conducted to relate them to more commonly available chemical properties such as solubility in water (S_w) and the octanol–water partition coefficient. The octanol–water partition coefficient is a dimensionless constant defined by

$$K_{ow} = \frac{C_{octanol}}{C_{water}} \quad \text{[Eq. II.3.12]}$$

where $C_{octanol}$ = concentration of an organic compound in octanol and C_{water} = concentration of the organic compound in water.

K_{ow} serves as an indicator of how an organic compound will partition between an organic phase and water. Values of K_{ow} range widely, from 10^{-3} to 10^7. Organic chemicals with low K_{ow} values are hydrophilic (likely to stay in water) and have low soil adsorption. There are many correlation equations between K_{oc} and K_{ow} (or solubility in water, S_w) reported in the literature. Table II.3.D lists the ones summarized in an EPA handbook.[5] It can be seen that K_{oc} increases linearly with increasing K_{ow} or with decreasing S_w on a log–log plot. (Note: Values of K_{ow} for some commonly found contaminants are provided in Table II.3.C.) The following simple correlation is also commonly used:[8]

$$K_{oc} = 0.63 K_{ow} \quad \text{[Eq. II.3.13]}$$

Table II.3.D Some Correlation Equations between K_{oc} and K_{ow}

Equation	Database
$\log K_{oc} = 0.544 (\log K_{ow}) + 1.377$ or $\log K_{oc} = -0.55 (\log S_w) + 3.64$	Aromatics, carboxylic acids and esters, insecticides, ureas and uracils, triazines, miscellaneous
$\log K_{oc} = 1.00 (\log K_{ow}) - 0.21$	Polycyclic aromatics, chlorinated hydrocarbons
$\log K_{oc} = -0.56 (\log S_w) + 0.93$	PCBs, pesticides, halogenated ethanes and propanes, PCE, 1,2-dichlorobenzene

Note: S_w is the solubility in water, in mg/L.

From U.S. EPA, *Site Characterization for Subsurface Remediation,* EPA 625/R-91/026, U.S. EPA, Washington, DC, 1991.

To estimate the solid concentration in equilibrium with the liquid concentration (or vice versa), we have to determine the value of the partition coefficient. The following procedure can be used to determine the partition coefficient for a soil–water system:

Step 1: Find K_{ow} or S_w for the compound of concern (Table II.3.C).
Step 2: Determine K_{oc} using correlations in Table II.3.D or Eq. II.3.13.
Step 3: Determine f_{oc} of the soil.
Step 4: Determine K_p from Eq. II.3.11.

Example II.3.3 Solid–liquid equilibrium concentrations

The aquifer underneath a site is contaminated with tetrachloroethylene (PCE). A groundwater sample contains 200 ppb of PCE. Estimate the PCE concentration adsorbed on the aquifer material, which contains 1% of organic carbon. Assume the adsorption follows a linear model.

Solution:

a. From Table II.3.C, for PCE

$$\text{Log } K_{ow} = 2.6 \rightarrow K_{ow} = 398$$

b. From Table II.3.D, for PCE, a chlorinated hydrocarbon

$$\text{Log } K_{oc} = 1.00(\text{Log } K_{ow}) - 0.21 = 2.6 - 0.21 = 2.39$$
$$K_{oc} = 245 \text{ mL/g} = 245 \text{ L/kg}$$

or, from Eq. II.3.13,

$$K_{oc} = 0.63 K_{ow} = 0.63(398) = 251 \text{ mL/g} = 251 \text{ L/kg}$$

c. Use Eq. II.3.11 to find K_p

$$K_p = f_{oc} K_{oc} = (1\%)(251) = 2.51 \text{ mL/g} = 2.51 \text{ L/kg}$$

d. Use Eq. II.3.10 to find X

$$X = K_p C = (2.51 \text{ L/kg})(0.2 \text{ mg/L}) = 0.50 \text{ mg/kg}$$

Discussion

1. The simple equation II.3.13 yields an estimate of K_{oc} comparable to that from the correlation equation in Table II.3.D.
2. Most books do not talk about the units of K_p, and even the correlation equations here do not mention them. Actually, K_p has a unit of "(volume of solvent)/(mass of adsorbent)," and it is equal to mL/g or L/kg in most, if not all, of the correlation equations.

II.3.4 Solid–liquid–vapor equilibrium

As mentioned at the beginning of this section, an NAPL may end up in four different phases as it enters a vadose zone. We have just discussed the equilibrium systems of liquid–vapor and soil–liquid. Now we move one step further to discuss the system including liquid, vapor, and solid (and free product in some of the applications).

The soil moisture in the vadose zone is in contact with both soil grains and air in the void, and the contaminant in each phase can travel to the other phases. The contaminant concentration in the liquid, for example, is affected by the concentrations in the other phases (i.e., soil, vapor, and free product). These concentrations are related by the equilibrium equations mentioned earlier. In other words, if the entire system is at equilibrium and the contaminant concentration of one phase is known, the concentrations at other phases can be estimated using the equilibrium relationships. Although in real applications, the equilibrium condition does not always exist; the estimate from such a condition serves as a good starting point or as the upper or the lower limit of the real values.

Example II.3.4 Solid–liquid–vapor equilibrium concentrations

Free-product phase of 1,1,1-trichloroethane (1,1,1-TCA) was found in the subsurface at a site. The soil is silty with an organic content of 2%. The subsurface temperature is 20°C. Estimate the maximum concentrations of TCA in the air void, in the liquid, and on the soil grain.

Solution:

a. Since free product is present, the maximum vapor concentration will be the vapor pressure of the TCA liquid at that temperature.

From Table II.3.C, the vapor pressure of TCA is 100 mmHg at 20°C.

$$100 \text{ mmHg} = (100 \text{ mmHg}) \div (760 \text{ mmHg}/1 \text{ atm})$$

$$= 0.132 \text{ atm}$$

$$G = 0.132 \text{ atm} = 132{,}000 \text{ ppmV}$$

Use Eq. II.1.1 to convert ppmV to mg/m^3 (molecular weight = 133.4 from Table II.3.C).

$$132{,}000 \text{ ppmV} = (132{,}000)[(133.4/24.05)] \text{ mg/m}^3$$

$$G = 732{,}200 \text{ mg/m}^3 = 732.2 \text{ mg/L}$$

b. From Table II.3.C, H = 14.4. Convert H to dimensionless Henry's constant, using Table II.3.B

$$H = H^*RT = 14.4 = H^* \,(0.082)(273 + 20)$$

$$H^* = 0.60 \text{ (dimensionless)}$$

Use Eq. II.3.5 to find the liquid concentration

$$G = HC = 732.2 \text{ mg/L} = (0.60)C$$

So,

$$C = 1220 \text{ mg/L} = 1220 \text{ ppm}$$

c. From Table II.3.C, for TCA

$$\text{Log } K_{ow} = 2.49 \rightarrow K_{ow} = 309$$

From Table II.3.D, for TCA, a chlorinated hydrocarbon

$$\text{Log } K_{oc} = 1.00(\text{Log } K_{ow}) - 0.21 = 2.49 - 0.21 = 2.28$$

$$K_{oc} = 191 \text{ mL/g} = 191 \text{ L/kg}$$

Or, from Eq. II.3.13,

$$K_{oc} = 0.63K_{ow} = 0.63(309) = 195 \text{ mL/g} = 195 \text{ L/kg}$$

Use Eq. II.3.11 to find K_p

$$K_p = f_{oc}\, K_{oc} = (2\%)(191) = 3.82 \text{ mL/g} = 3.82 \text{ L/kg}$$

Use Eq. II.3.10 to find the soil concentration, X

$$X = K_pC = (3.82 \text{ L/kg})(1220 \text{ mg/L}) = 4660 \text{ mg/kg}$$

Discussion

1. The calculated liquid concentration, 1220 mg/L, is lower than the solubility, 4400 mg/L given in Table II.3.C.
2. The simple equation, Eq. II.3.13, yields an estimate of K_{oc} comparable to that from the correlation equation in Table II.3.D.
3. The calculated concentrations are the maximum possible values; the actual values would be lower if the system is not at equilibrium.

II.3.5 Partition of contaminants in different phases

The total mass of contaminants in the vadose zone is the sum of the mass in four phases. Let us consider a contaminant plume in the vadose zone with a volume, V.

From Eq. II.1.3,

$$\text{mass of contaminants in the soil moisture} = (V_l)(C) = [V\phi_w]C \qquad \text{[Eq. II.3.14]}$$

From Eq. II.1.4,

$$\text{mass of contaminants adsorbed to the soil grains} = (M_S)(X) = [V(\rho_b)]X \qquad \text{[Eq. II.3.15]}$$

From Eq. II.1.5,

$$\text{mass of contaminants in the pore void space} = (V_a)(G) = [(V)(\phi_a)]G \qquad \text{[Eq. II.3.16]}$$

where ϕ_w is the volumetric water content and ϕ_a is the air porosity (total porosity, $\phi = \phi_w + \phi_a$). The total mass of contaminant, M_t, present in the plume is the sum of the above three and free product, if any. Thus,

$$M_t = V(\phi_w)C + V(\rho_b)X + V(\phi_a)G + \text{free product} \qquad \text{[Eq. II.3.17]}$$

The mass of free product is simply the volume of the free product multiplied by its mass density. If no free product is present, Eq. II.3.17 can be simplified to

$$M_t = V(\phi_w)C + V(\rho_b)X + (V)(\phi_a)G \qquad \text{[Eq. II.3.18]}$$

If the system is in equilibrium and Henry's law and linear adsorption apply, the concentration in one phase can be represented by the concentration in another phase multiplied by a factor. The following relationships exist:

$$G = HC = H\left(\frac{X}{K_p}\right) = \left(\frac{H}{K_p}\right)X \qquad \text{[Eq. II.3.19]}$$

$$C = \left(\frac{X}{K_p}\right) = \left(\frac{G}{H}\right) \qquad \text{[Eq. II.3.20]}$$

$$X = K_pC = K_p\left(\frac{G}{H}\right) = \left(\frac{K_p}{H}\right)G \qquad \text{[Eq. II.3.21]}$$

Combining the above relationships and Eq. II.3.18, Eq. II.3.18 can be rearranged as

$$\begin{aligned} \frac{M_t}{V} &= [(\phi_w) + (\rho_b)K_p + (\phi_a)H]C \\ &= \left[\frac{(\phi_w)}{H} + \frac{(\rho_b)K_p}{H} + (\phi_a)\right]G \qquad \text{[Eq.II.3.22]} \\ &= \left[\frac{(\phi_w)}{K_p} + \rho_b + (\phi_a)\frac{H}{K_p}\right]X \end{aligned}$$

where M_t/V can be viewed as the average mass concentration of the plume. The total mass of contaminants in a plume can be readily determined by multiplying (M_t/V), if known, with the total volume of the plume. Equation II.3.22 can be used to estimate the total mass of contaminant in a vadose zone, if the average liquid concentration, soil concentration, or vapor concentration is known, when no free product is present.

For a dissolved groundwater plume ($\phi_a = 0$ and $\phi_w = \phi$), Eq. II.3.22 can be modified as

$$\frac{M_t}{V} = [\phi + (\rho_b)K_p]C$$

$$= \left[\frac{\phi}{K_p} + \rho_b\right]X \qquad \text{[Eq. II.3.23]}$$

To use the equations in this subsection, the following units are suggested: V (in liters), G (mg/L), C (mg/L), X (mg/kg), M_t (mg), ρ_b (kg/L), K_p (L/kg), and ϕ, ϕ_w, ϕ_a, and H (dimensionless).

Example II.3.5A Mass partition between vapor and liquid phase
A new field technician was sent out to collect a groundwater sample from a monitoring well. He filled only half of the 40-mL sample vial with groundwater contaminated with benzene (T = 20°C). The benzene concentration in the collected groundwater was analyzed to be 5 mg/L.

Determine

a. The concentration of benzene in the headspace (in ppmV) before the vial was opened.
b. The percentage of total benzene mass in the aqueous phase of the closed vial.
c. The true benzene concentration in the groundwater, if headspace free sample is collected.

Assume the value of the dimensionless Henry's constant for benzene is 0.22.

Solution:
Basis: 1-L container

a. Concentration of benzene in the headspace:

$$H \times C_l = (0.22)(5) = 1.1 \text{ mg/L} = 1100 \text{ mg/m}^3$$

1 ppmV = (MW/24.05) mg/m³ = (78/24.05) mg/m³ = 3.24 mg/m³

Concentration of benzene in the headspace = 1100/3.24 = 340 ppmV

b. Mass of benzene in the liquid phase:

$$(C)(V) = (5)(0.5) = 2.5 \text{ mg}$$

Mass of benzene in the headspace:

$$(G)(V) = (1.1)(0.5) = 0.55 \text{ mg}$$

Total mass of benzene:

$$\text{mass in liquid} + \text{mass in headspace} = 2.5 + 0.55 = 3.05 \text{ mg}$$

Percentage of total benzene mass in the aqueous phase = 2.5/3.05 = 82%.

c. The actual liquid concentration should be

$$(3.05)/(0.5) = 6.1 \text{ mg/L}$$

Discussion

1. Although the sample volume is only 40 mL, the calculation basis was 1 L to simplify the calculation.
2. With headspace in the sample bottle, the apparent liquid concentration was lower than the actual concentration.

Example II.3.5B Mass partition between solid and liquid phase

The aquifer underneath a site is contaminated with tetrachloroethylene (PCE). The aquifer porosity is 0.4, and the bulk density of the aquifer material is 1.8 g/cm^3. A groundwater sample contains 200 ppb of PCE.

Assuming that the adsorption follows a linear model, estimate:

a. The PCE concentration adsorbed on the aquifer material, which contains 1% by weight of organic carbon.
b. The partition of PCE in the two phases, i.e., dissolved phase and adsorbed onto the solid phase.

Solution:

a. The PCE concentration adsorbed onto the solid has been determined in Example II.3.3 as 0.50 mg/kg.

b. Basis: 1-L aquifer formation.
Mass of PCE in the liquid phase:

$$(C)[(V)(\phi)] = (0.2)[(1)(0.4)] = 0.08 \text{ mg}$$

Mass of PCE adsorbed on the solid:

$$(X)[(V)(\rho_b)] = (0.5)[(1)(1.8)] = 0.9 \text{ mg}$$

Total mass of PCE:

mass in liquid + mass on the solid = 0.08 + 0.9 = 0.98 mg

Percentage of total PCE mass in the aqueous phase = 0.08/0.98 = 8.2%

Discussion. Most of the PCE, 91.8%, in the contaminated aquifer is adsorbed onto the aquifer materials. This partially explains why the clean-up of aquifer takes a long time using the pump-and-treat method.

Example II.3.5C Mass partition between liquid and solid phase
A wastewater contains 500 mg/L of suspended solids. The fraction of organics of the solids is 1% by weight. The benzene concentration of the *filtered* wastewater is determined to be 5 mg/L. The K_{oc} of benzene is 83 mL/g.
Determine

a. The concentration of benzene in suspended solids.
b. The percentage of total benzene mass in the dissolved phase of the unfiltered wastewater.

Solution:

a. Use Eq. II.3.11 to find K_p.

$$K_p = f_{oc}K_{oc} = (1\%)(83) = 0.83 \text{ mL/g} = 0.83 \text{ L/kg}$$

Use Eq. II.3.10 to find X.

$$X = K_pC = (0.83 \text{ L/kg})(5 \text{ mg/L}) = 4.15 \text{ mg/kg}$$

b. Basis: 1-L solution.
Mass of benzene in the liquid phase:

$$(C)[(V)] = (5)(1) = 5 \text{ mg}$$

Mass of benzene adsorbed on the solid:

$$(X)[(V)(\text{suspended solid concentration})]$$
$$= (4.15 \text{ mg/kg})[(1 \text{ L})(5000 \text{ mg/L})(1 \text{ kg}/1{,}000{,}000 \text{ mg}]$$
$$= 2.075 \times 10^{-3} \text{ mg}$$

Total mass of benzene:

mass in liquid + mass on the solid = 5 + 2.075 × 10^{-3} = 5.0021 mg

Percentage of total benzene mass in the aqueous phase = 5/5.0021 = 99.96%.

Discussion. Almost all of the benzene, 99.96%, is in the dissolved phase because only a small amount of solids is present.

Example II.3.5D *Mass partition among vapor, liquid, and solid phases*

The vapor concentrations of benzene and pyrene in the void space of the vadose zone underneath a landfill are 100 ppmV and 10 ppbV, respectively. The porosity of the vadose zone is 40%, and 30% of the porosity is occupied by water. The bulk density of the soil is 1.8 g/cm^3. Assuming no free product phase is present, determine the mass fractions of each compound in the three phases, i.e., void space, moisture, and solid phases. The values of the dimensionless Henry's constant for benzene and pyrene are 0.22 and 0.0002, respectively. The values of the K_p for benzene and pyrene are 1.28 and 1446, respectively.

Strategy. Using a computer spreadsheet, such as EXCEL, is a good way to solve a problem such as this.

Solution:

Basis: 1 m^3 of soil.

		Benzene	Pyrene
a.	Determine the mass in the air void		
	Molecular weight	78	202
	G (ppmV)	100	0.01
	G (mg/m^3)	324.32	0.08
	Air void (m^3)= 0.40 * 0.7	0.28	0.28
	Mass in void (mg)	90.81	0.024
b.	Determine the mass dissolved in the liquid		
	H	0.22	0.0002
	C (mg/m^3) = G/H	1474.2	420.0
	Liq. vol(m^3) = 0.40 * 0.3	0.12	0.12
	Mass in liquid (mg)	176.9	50.4
c.	Determine the mass attached onto the solid		
	K_p	1.28	717
	C (mg/L)	1.47	0.4
	X(mg/kg) = $K_p \times$ C(mg/L)	1.9	286.8
	Soil mass (kg) = (1 m^3)(ρ_b)	1800	1800
	Mass in solid (mg)	3396.6	5.16×10^5

		Benzene	Pyrene
d.	Determine the total mass in three phases		
	Total pollutant (mg)	3664.3	5.16×10^5
e.	Determine the mass fraction in each phase		
	% in void	2.5	4.6×10^{-6}
	% in moisture	4.8	9.7×10^{-3}
	% in solid	92.7	1.0×10^2

Discussion. For both compounds, most of the contaminants are attached onto the solid. This is especially true for pyrene, which has very high K_p and low H values. The vapor concentration of pyrene is extremely low, but the soil concentration is very high.

Example II.3.5E *Relationship between soil vapor concentration and soil sample concentration*

The vapor concentrations of benzene and pyrene in the void space of the vadose zone underneath a landfill are 100 ppmV and 10 ppbV, respectively, from a soil gas survey. The porosity of the vadose zone is 40%, 30% of the porosity is occupied by water, and the bulk density of the soil is 1.8 g/cm^3. The values of the dimensionless Henry's constant for benzene and pyrene are 0.22 and 0.0002, respectively. The values of the K_p for benzene and pyrene are 1.28 and 1446, respectively.

Soil samples were taken from the location where the soil gas probe was located and analyzed in a laboratory for the contaminant concentrations in soil. Estimate the contaminant concentrations in soil.

Solution:

Basis: 1 L of soil.

a. Let us work on benzene first. We have to convert the vapor concentration in ppmV into mg/L. From Example II.3.5D, G = 0.324 mg/L for benzene. Use Eq. II.3.22 to estimate the soil concentration of benzene:

$$\frac{M_t}{V} = \left[\frac{(\phi_w)}{H} + \frac{(\rho_b)K_p}{H} + (\phi_a)\right]G \qquad \text{[Eq.II.3.22]}$$

$$= \left[\frac{0.12}{0.22} + \frac{(1.8)(1.28)}{0.22} + 0.28\right](0.324) = 3.63 \text{ mg/L}$$

To convert the soil concentration into mg/kg, we should divide the value by the bulk density of the soil:

Soil concentration = 3.63 mg/L ÷ 1.8 kg/L = 2.0 mg/kg

b. For pyrene, from Example II.3.5D, G = 0.000084 mg/L. Use Eq. II.3.22 to estimate the soil concentration of pyrene:

$$\frac{M_t}{V} = \left[\frac{(\phi_w)}{H} + \frac{(\rho_b)K_p}{H} + (\phi_a)\right]G$$

$$= \left[\frac{0.12}{0.0002} + \frac{(1.8)(1446)}{0.0002} + 0.28\right](0.000084) \qquad \text{[Eq.II.3.23]}$$

$$= 1093 \text{ mg/L}$$

To convert the soil concentration into mg/kg, we should divide the value by the bulk density of the soil:

Soil concentration = 1093 mg/L ÷ 1.8 kg/L = 607 mg/kg

Discussion

1. In this example, a soil sample containing 2 mg/kg benzene yields a soil vapor concentration of 100 ppmV. The soil concentration of pyrene, 607 mg/kg, is 300 times higher than that of benzene, but its vapor concentration is much lower.
2. For a given contaminant concentration in soil, its soil vapor concentration will be higher if K_p value is smaller and the Henry's constant of the contaminant is larger. (In other words, the soil contains fewer organics, and the contaminant is less hydrophobic and more volatile.) For sandy soil, the soil vapor concentration may be high, but the mass adsorbed to the sand grains can be relatively low. This explains why PID or OVA readings on contaminated sandy soil samples may be high; however, the laboratory results on contaminant concentrations of sandy soil turned out to be low.

References

1. Ballestero, T. P., Fiedler, F. R., and Kinner, N. E., An investigation of the relationship between actual and apparent gasoline thickness in a uniform sand aquifer, *Groundwater,* 32, 708, 1994.
2. Driscoll, F. G., *Groundwater and Wells,* 2nd ed., Johnson Division, St. Paul, MN, 1986.
3. U.S. EPA, *Transport and Fate of Contaminants in the Subsurface,* EPA 625/4-89/019, U.S. EPA, Washington, DC, 1989.
4. U.S. EPA, *CERCLA Site Discharges to POTWs Treatability Manual,* EPA 540/2-90-007, Office of Water, U.S. EPA, Washington, DC, 1990.

5. U.S. EPA, *Site Characterization for Subsurface Remediation*, EPA 625/R-91/026, U.S. EPA, Washington, DC, 1991.
6. Fetter, C. W., Jr., *Applied Hydrogeology*, Charles E. Merrill Publishing, Columbus, OH, 1980.
7. Kuo, J. F. and Cordery, S. A., Discussion of nomograph for air stripping of VOC from water, *J. Environ. Eng.*, V. 114, No. 5, p. 1248–1250, 1988.
8. LaGrega, M. D., Buckingham, P. L., and Evans, J. C., *Hazardous Waste Management*, McGraw-Hill, New York, 1994.
9. Lide, D. R., *Handbook of Chemistry and Physics*, 73rd ed., CRC Press, Boca Raton, FL, 1992.
10. Reid, R. C., Prausnitz, J. M., and Poling, B. F., *The Properties of Liquids and Gases*, 4th ed., McGraw-Hill, New York, 1987.
11. Todd, D. K., *Groundwater Hydrology*, 2nd ed., John Wiley & Sons, New York, 1980.

chapter three

Plume migration in groundwater and soil

In Chapter two we illustrated the necessary calculations for site characterization and remedial investigation. Generally, from the RI activities the extent of the plume in the vadose zone and/or groundwater is defined. If the contaminants cannot be removed immediately, they will migrate under common field conditions and the extent of the plume will enlarge.

In the vadose zone, the contaminants will move downward as a free product or become dissolved in infiltrating water and then move downward by gravity. The downward-moving liquid may come in contact with the underlying aquifer and create a dissolved plume. In addition, the VOCs will volatilize into the air void of the vadose zone and travel under advective forces (with the air flow) or concentration gradients (through diffusion). Migration of the vapor can be in any direction, and the contaminants in the vapor phase, when coming in contact with the groundwater, may also dissolve into the groundwater. For site remediation or health risk assessment, understanding the fate and transport of contaminants in the subsurface is important. Common questions related to the fate and transport of contaminants in the subsurface include

1. How long will it take for the plume in the vadose zone to enter the aquifer?
2. How far will the vapor contaminants in the vadose zone travel? In what concentrations?
3. How fast does the groundwater flow? In which direction?
4. How fast will the plume migrate? In which direction?
5. Will the plume migrate at the same speed as the groundwater flow or at a different speed? If different, what are the factors that would

make the plume migrate at a different speed from the groundwater flow?

6. How long has the plume been present in the aquifer?

This chapter illustrates the basic calculations needed to answer most of the above questions. The first section presents the calculations for groundwater movement and clarifies some common misconceptions about groundwater velocity and hydraulic conductivity. Procedures to determine the groundwater flow gradient and the flow direction are also given. The second section presents groundwater extraction from confined and unconfined aquifers. Since hydraulic conductivity plays a pivotal role in groundwater movement, several common methodologies of estimating this parameter are covered, including the aquifer tests. The discussion then moves to the migration of the dissolved plume in the aquifer and in the vadose zone.

III.1 Groundwater movement

III.1.1 Darcy's law

Darcy's Law is commonly used to describe laminar flow in porous media. For a given medium the flow rate is proportional to the head loss and inversely proportional to the length of flow path. Flow in typical groundwater aquifers is laminar, and therefore Darcy's Law is valid. Darcy's Law can be expressed as

$$v = \frac{Q}{A} = -K\frac{dh}{dl} \qquad \text{[Eq. III.1.1]}$$

where v is the Darcy velocity, Q is the volumetric flow rate, A is the cross-sectional area of the porous medium perpendicular to the flow, dh/dl is the hydraulic gradient (a dimensionless quantity), and K is the hydraulic conductivity.

The hydraulic conductivity tells how permeable the porous medium is to the flowing fluid. The larger the K of a formation, the easier the fluid flows through it.

Commonly used units for hydraulic conductivity are either in velocity units such as ft/d, cm/s, or m/d, or in volumetric flow rate per unit area such as gpd/ft^2. You may find the unit conversions in Table III.1.A helpful.

Example III.1.1 *Estimate the rate of fresh groundwater in contact with the plume*

Leachates from a landfill leaked into the underlying aquifer and created a contaminated plume. Use the information below to estimate the amount of fresh groundwater that enters into the contaminated zone per day.

The maximum cross-sectional area of the plume perpendicular to the groundwater flow = 1600 ft^2
Groundwater gradient = 0.005
Hydraulic conductivity = 2500 gpd/ft^2

Solution:

Another common form of Darcy's Law (Eq. III.1.1) is

$$Q = KiA \qquad \text{[Eq. III.1.2]}$$

where i is the hydraulic gradient, dh/dl.

The rate of fresh groundwater entering the plume can be found by inserting the appropriate values into the above equation:

$$Q = (2500\ \text{gpd/ft}^2)(0.005)(1600\ \text{ft}^2) = 20{,}000\ \text{gpd}$$

Discussion

1. The calculation itself is straightforward and simple. However, we can get valuable and useful information from this exercise. The rate of 20,000 gal/day represents the rate of uncontaminated groundwater that will come in contact with the contaminants. This water would become contaminated and move downstream or sidestream and, consequently, enlarge the size of the plume. To control the spread of the plume, we have to extract this amount of water, 20,000 gpd or ~14 gpm, as a minimum. The actual extraction rate required should be higher than this because the groundwater drawdown from pumping will increase the flow gradient. This increased gradient will, in turn, increase the rate of groundwater entering the plume zone as indicated by the equation above.
2. Using the maximum cross-sectional area is a legitimate approach that represents the "contact face" between the fresh groundwater and the plume.

III.1.2 *Darcy's velocity vs. seepage velocity*

The velocity term in Eq. III.1.1 is called the Darcy velocity (or the discharge velocity). Does this Darcy velocity represent the groundwater flow velocity?

Table III.1.A Common Conversion Factors for Hydraulic Conductivity

m/d	cm/s	ft/d	gpd/ft²
1	1.16E − 3	3.28	2.45E + 1
8.64E + 2	1	2.83E + 3	2.12E + 4
3.05E − 1	3.53E − 4	1	7.48
4.1E − 2	4.73E − 5	1.34E − 1	1

The answer is "no." The Darcy velocity in that equation assumes the flow occurs through the entire cross-section of the porous medium. In other words, it is the velocity at which water would move through an aquifer if the aquifer were an open conduit. Actually, the flow is limited to the available pore space only (the effective cross-sectional area available for flow is smaller), so the actual fluid velocity through the porous medium would be larger than the Darcy velocity. This flow velocity is often called the seepage velocity or the interstitial velocity. The relationship between the seepage velocity, v_s, and the Darcy velocity, v, is as follows:

$$v_s = \frac{Q}{\phi A} = \frac{v}{\phi} \qquad \text{[Eq. III.1.3]}$$

where ϕ is the porosity. For example, for an aquifer with a porosity of 33%, the seepage velocity of groundwater flowing through this aquifer will be three times the Darcy velocity (i.e., $v_s = 3\ v$).

Example III.1.2 *Determine Darcy velocity and seepage velocity*

There is spill of an inert (or a conservative) substance into the subsurface. The spill infiltrates the unsaturated zone and quickly reaches the underlying water table aquifer. The aquifer consists mainly of sand and gravel with a hydraulic conductivity of 2500 gpd/ft^2 and an effective porosity of 0.35. The water level in a well neighboring the spill lies at an altitude of 560 ft, and the level in another well 1 mile directly down gradient is 550 ft. Determine

a. The Darcy velocity of the groundwater
b. The seepage velocity of the groundwater
c. The velocity of plume migration
d. How long it will take for the plume to reach the down-gradient well

Solution:

a. We have to determine the gradient of the aquifer first:

$$i = dh/dl = (560 - 550)/5280 = 1.89 \times 10^{-3} \text{ ft/ft}$$

Darcy velocity = Ki

$$\left[(2500\ \text{gpd/ft}^2)\left(0.134\frac{\text{ft/d}}{\text{gpd/ft}^2}\right)\right](1.89\times10^{-3}\ \text{ft/ft}) = 0.63\ \text{ft/d}$$

b. Seepage velocity = v/ϕ

$$0.63/0.35 = 1.81\ \text{ft/d}$$

c. The pollutant is inert, meaning that it will not react with the aquifer. (Sodium chloride is a good example of an inert substance and is often used as a tracer in an aquifer study.) Therefore, the velocity of plume migration for this case is the same as the seepage velocity, 1.81 ft/d.
d. Time = distance/velocity

$$5280 \text{ ft}/(1.81 \text{ ft/d}) = 2912 \text{ days} = 8.0 \text{ year}$$

Discussion

1. The conversion factor, 1 gpd/ft^2 = 0.134 ft/d, used in (a) is from Table III.1.A.
2. The calculated plume migration velocity is crude at best and should only be considered as a rough estimate. Many factors, such as hydrodynamic dispersion, are not considered in this equation. The dispersion can cause parcels of water to spread transversely to the major direction of groundwater flow and move longitudinally, down gradient, at a faster rate. The dispersion is caused by an intermixing of water particles due to the differences in interstitial velocity induced by the heterogeneous pore sizes and tortuosity.
3. In addition, the migration of most chemicals will be retarded by interactions with the geologic formation, especially with clays, soil–organic matter, and metal oxides and hydroxides. This phenomenon will be discussed further in Section III.4.3.

III.1.3 *Intrinsic permeability vs. hydraulic conductivity*

In the soil venting literature one may encounter a statement such as "the soil permeability is 4 Darcies," while in groundwater remediation literature one may read that "the hydraulic conductivity is equal to 3 cm/s." Both statements describe how permeable the formations are. Are they the same? If not, what is the relationship between the permeability and hydraulic conductivity?

These two terms, permeability and hydraulic conductivity, are sometimes used interchangeably. However, they do have different meanings. The intrinsic permeability of a porous medium (i.e., a rock or soil) defines its ability to transmit a fluid. It is a property of the medium only and is independent of the properties of the transmitting fluid. That is why it is called the "intrinsic" permeability. On the other hand, the hydraulic conductivity of a porous medium depends on the properties of the fluid flowing through it.

Hydraulic conductivity is conveniently used to describe the ability of an aquifer to transmit groundwater. A porous medium has a unit hydraulic conductivity if it will transmit a unit volume of groundwater through a unit cross-sectional area (perpendicular to the direction of flow) in a unit time at the prevailing kinematic viscosity and under a unit hydraulic gradient.

The relationship between the intrinsic permeability and hydraulic conductivity is

$$K = \frac{k\rho g}{\mu} \quad \text{or} \quad k = \frac{K\mu}{\rho g} \qquad \text{[Eq. III.1.4]}$$

where K is the hydraulic conductivity, k is the intrinsic permeability, μ is the fluid viscosity, ρ is the fluid density, and g is the gravitational constant (kinematic viscosity = μ/ρ). The intrinsic permeability has a unit of area as shown below:

$$k = \frac{K\mu}{\rho g} = \left[\frac{(\text{m/s})(\text{kg/m}\cdot\text{s})}{(\text{kg/m}^3)(\text{m/s}^2)}\right] = [\text{m}^2] \qquad \text{[Eq. III.1.5]}$$

In petroleum industries the intrinsic permeability of a formation is measured by a unit termed Darcy. A formation has an intrinsic permeability of 1 Darcy if it can transmit a flow of 1 cm^3/s with a viscosity of 1 centipoise under a pressure gradient of 1 atmosphere/cm, that is,

$$1 \text{ Darcy} = \frac{\dfrac{(1 \text{ g/cm}\cdot\text{s})(1 \text{ cm}^3/\text{s})}{1 \text{ cm}^2}}{1 \text{ atmosphere/cm}} \qquad \text{[Eq. III.1.6]}$$

By substitution of appropriate units, it can be shown that

$$1 \text{ Darcy} = 0.987 \times 10^{-8} \text{cm}^2 \qquad \text{[Eq. III.1.7]}$$

Table III.1.B lists the mass density and viscosity of water under one atmosphere. As shown in the table, the density of water from 0 to 30°C is essentially the same, at 1 g/cm^3; the viscosity of water decreases with increasing temperature. The viscosity of water at 20°C is one centipoise. (This is the viscosity value of the fluid used in defining the Darcy unit.)

Example III.1.3 *Determine hydraulic conductivity from a given intrinsic permeability*

The intrinsic permeability of a soil core sample is 1 Darcy. What is the hydraulic conductivity of this soil for water at 15°C? How about at 25°C?

Solution:

a. At 15°C, density of water (15°C) = 0.999703 g/cm^3 (from Table III.1.B), and viscosity of water (15°C) = 0.01139 poise = 0.01139 $\text{g/s}\cdot\text{cm}$ (from Table III.1.B).

$$K = \frac{k\rho g}{\mu} = \frac{(9.87 \times 10^{-9}\ \text{cm}^2)(0.999703\ \text{g/cm}^3)(980\ \text{cm/s}^2)}{0.01139\ \text{g/s} \cdot \text{cm}}$$

$$= 8.49 \times 10^{-4}\ \text{cm/s}$$

$$= (8.49 \times 10^{-4})(2.12 \times 10^4) = 18.0\ \ \text{gpd/ft}^2 = 18.0\ \text{meinzers}$$

b. At 25°C, density of water (25°C) = 0.997048 g/cm^3 (from Table III.1.B), and viscosity of water (25°C) = 0.00890 poise = 0.00890 g/s · cm (from Table III.1.B).

$$K = \frac{k\rho g}{\mu} = \frac{(9.87 \times 10^{-9}\ \text{cm}^2)(0.999703\ \text{g/cm}^3)(980\ \text{cm/s}^2)}{0.00890\ \text{g/s} \cdot \text{cm}}$$

$$= 1.09 \times 10^{-3}\ \text{cm/s}$$

$$= (1.09 \times 10^{-3})(2.12 \times 10^4) = 23.0\ \ \text{gpd/ft}^2$$

Discussion. This example illustrates that a porous medium with an intrinsic permeability of 1 Darcy has a hydraulic conductivity of 18 gpd/ft^2 at 15°C (23 gpd/ft^2 at 25°C). The unit of gpd/ft^2 is commonly used by hydrogeologists in the United States. The unit is also named the meinzer after O. E. Meinzer, a pioneering groundwater hydrogeologist with U.S. Geological Services.[2] The unit of cm/s is more commonly used in soil mechanics. (For example, the hydraulic conductivity of clay liners or flexible membrane liners in landfills is commonly expressed in cm/s.)

From the above example, one can tell that a geologic formation with an intrinsic permeability of one Darcy has a hydraulic conductivity of approximately 10^{-3} cm/s or 20 gpd/ft^2 for transmitting pure water at 20°C. Typical

Table III.1.B Physical Properties of Water under One Atmosphere

Temperature (°C)	Density (g/cm^3)	Viscosity (cp)
0	0.999842	1.787
3.98	1.000000	1.567
5	0.999967	1.519
10	0.999703	1.307
15	0.999103	1.139
20	0.998207	1.002
25	0.997048	0.890
30	0.995650	0.798
40	0.992219	0.653

Note: 1 g/cm^3 = 1000 kg/m^3 = 62.4 lb/ft^3. 1 centipoise = 0.01 poise = 0.01 g/cm · s = 0.001 Pa · s = 2.1 × 10^{-5} lb · s/ft^2.

Table III.1.C Typical Values of Intrinsic Permeabilities and Hydraulic Conductivities

	Intrinsic permeability (Darcy)	Hydraulic conductivity (cm/s)	Hydraulic conductivity (gpd/ft^2)
Clay	10^{-6}–10^{-3}	10^{-9}–10^{-6}	10^{-5}–10^{-2}
Silt	10^{-3}–10^{-1}	10^{-6}–10^{-4}	10^{-2}–1
Silty sands	10^{-2}–1	10^{-5}–10^{-3}	10^{-1}–10
Sands	1–10^{2}	10^{-3}–10^{-1}	10–10^{3}
Gravel	10–10^{3}	10^{-2}–1	10^{2}–10^{4}

values of intrinsic permeabilities and hydraulic conductivities for different types of formations are given in Table III.1.C.

III.1.4 Transmissivity, specific yield, and storativity

Transmissivity (T) is another concept that is commonly used to describe an aquifer's capacity to transmit water. It represents the amount of water that can be transmitted horizontally by the entire saturated thickness of the aquifer under a hydraulic gradient of one. It is equal to the multiplication product of the aquifer thickness (b) and the hydraulic conductivity (K). Commonly used units for T are m^2/d and gpd/ft.

$$T = Kb \qquad \text{[Eq. III.1.8]}$$

An aquifer typically serves two functions: (1) a conduit through which flow occurs and (2) a storage reservoir. This is accomplished by the openings in the aquifer matrix. If a unit of saturated formation is allowed to drain by gravity, not all of the water it contains will be released. The ratio of water that can be drained by gravity to the entire volume of a saturated soil is called specific yield, while the part retained is the specific retention. Table III.1.D lists typical porosity, specific yield, and specific retention of soil, clay, sand, and gravel. The sum of the specific yield and the specific retention of a formation is equal to its porosity.

The specific yield and the specific retention are related to the attraction between water and the formation materials. Clayey formations usually have a lower hydraulic conductivity. This often leads to an incorrect idea that clayey formations have a lower porosity. As shown in Table III.1.D, clay has a much higher porosity than sand, and sand has a higher porosity than gravel. The porosity of clay can be as high as 50%, but its specific yield is extremely low at 2%. Porosity determines the total volume of water that a formation can store, while specific yield defines the amount that is available to pumping. The low specific yield explains the difficulty of extracting groundwater from clayey aquifers.

When the head in a saturated aquifer changes, water will be taken into or released from storage. Storativity or storage coefficient (S) describes the

Table III.1.D Typical Porosity, Specific Yield, and Specific Retention of Selected Materials

	Porosity (%)	Specific yield (%)	Specific retention (%)
Soil	55	40	15
Clay	50	2	48
Sands	25	22	3
Gravel	20	19	1

From U.S. EPA, *Ground Water Volume I: Ground Water and Contamination*, EPA/625/6-90/016a, U.S. EPA, Washington, DC, 1990.

quantity of water taken into or released from storage per unit change in head per unit area. It is a dimensionless quantity. The response of a confined aquifer to the change of water head is different from that of an unconfined aquifer. When the head declines, a confined aquifer remains saturated; the water is released from storage by the expansion of water and compaction of aquifer. The amount of release is exceedingly small. On the other hand, the water table rises or falls with change of head in an unconfined aquifer. As the water level changes, water drains from or enters into the pore spaces. This storage or release is mainly due to the specific yield. It is also a dimensionless quantity. For unconfined aquifers the storativity is practically equal to the specific yield and ranges typically between 0.1 and 0.3. The storativity of confined aquifers is substantially smaller and generally ranges between 0.0001 and 0.00001, and that for leaky confined aquifers is in the range of 0.001. A small storativity implies that it will require a larger pressure change (or gradient) to extract groundwater at a specific flow rate.[7]

The volume of groundwater (V) drained from an aquifer can be determined from the following:

$$V = SA(\Delta h) \qquad \text{[Eq. III.1.9]}$$

where S is the storativity, A is the area of the aquifer, and Δh is the change in head.

Example III.1.4 *Estimate loss of storage in aquifers due to change of head*

An unconfined aquifer has an area of 5 square miles. The storativity of this aquifer is 0.15. The water table falls 0.8 feet during a drought. Estimate the amount of water lost from storage.

If the aquifer is confined and its storativity is 0.0005, what would be the amount lost for a decrease of 0.8 feet in head?

Solution:

a. Inserting the values into Eq. III.1.9, we obtain the volume of water drained for the unconfined aquifer:

$$V = (0.15)[(5)(5280)^2 \text{ ft}^2](0.8 \text{ ft}) = 1.67 \times 10^7 \text{ ft}^3 = 1.25 \times 10^8 \text{ gal}$$

b. For the confined aquifer:

$$V = (0.0005)[(5)(5280)^2 \text{ ft}^2](0.8 \text{ ft}) = 5.58 \times 10^4 \text{ ft}^3 = 4.17 \times 10^5 \text{ gal}$$

Discussion. For the same amount of change in head, the water lost in the unconfined aquifer is 300 times more, which is the ratio of the two storativity values (0.15/0.0005 = 300).

III.1.5 Determine groundwater flow gradient and flow direction

Having a good knowledge of the gradient and direction of groundwater flow is vital to groundwater remediation. The gradient and the direction of flow have great impacts on selection of remediation schemes to control plume migration, such as location of the pumping wells and groundwater extraction rates, etc.

Estimates of the gradient and direction of groundwater flow can be made with a minimum of three groundwater elevations. The general procedure is described below and an example follows.

Step 1: Locate the three surveyed points on a map to scale.
Step 2: Connect the three points and mark their water table elevations on the map.
Step 3: Subdivide each side of the triangle into a number of segments of equal size. (Each segment represents an increment of elevation.)
Step 4: Connect the points of equal values of elevation (equipotential lines), which then form the groundwater contours.
Step 5: Draw a line that passes through and is perpendicular to each equipotential line. This line marks direction of flow.
Step 6: Calculate the groundwater gradient from the formula, $i = dh/dl$.

Example III.1.5 Estimate the gradient and direction of groundwater flow from three groundwater elevations

Three groundwater monitoring wells were installed at a contaminated site. Groundwater elevations were determined from a recent survey of these wells and the values were marked on a map. Estimate the flow gradient and direction of the groundwater flow in the underlying aquifer.

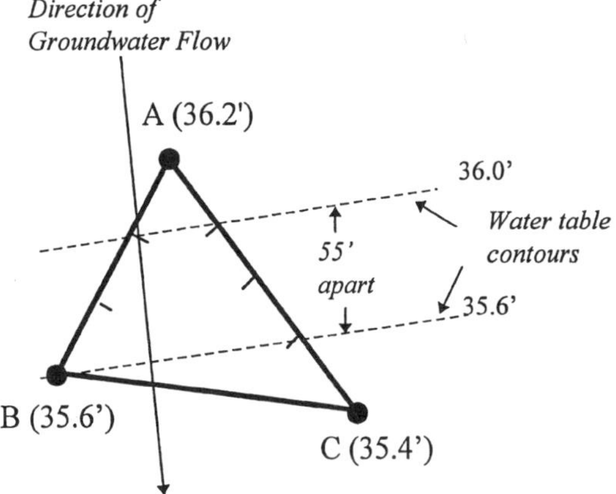

Figure E.III.1.5 Determination of groundwater gradient and direction.

Solution:

a. Water elevations (36.2′, 35.6′, and 35.4′) were measured at three monitoring wells and marked on the map.
b. These three points are connected by straight lines to form a triangle.
c. Subdivide each side of the triangle into a number of segments of equal intervals. For example, subdivide the line connecting point A (36.2′) and point B (35.6′) into three intervals. Each interval represents a 0.2′ increment in elevation.
d. Connect the points of equal values of elevation (equipotential lines), which then form the groundwater contours. Here, we connect the elevations of 36.0′ and 35.6′ to form two contour lines.
e. Draw a line that passes through and is perpendicular to each equipotential line and mark it as the groundwater flow direction.
f. Measure the distance between two contour lines, 55 feet in this example. Calculate the groundwater gradient from the formula, $i = dh/dl$:

$$i = (36.0 - 35.6)/(55) = 0.0073$$

Discussion. The groundwater elevations, especially those of the water table aquifers, may change with time. Consequently, the groundwater flow gradient and direction would change. Periodic surveys of the groundwater elevation may be necessary if fluctuation of the water table is suspected. Off-site pumping, seasonal change, and recharge are some of the reasons that may cause the fluctuation of the water table elevation.

III.2 Groundwater pumping

III.2.1 Steady-state flow in a confined aquifer

The equation describing steady-state flow of a confined aquifer (an artesian aquifer) from a fully penetrating well is shown below. A fully penetrating well means that the groundwater can enter at any level from the top to the bottom of the aquifer.

$$Q = \frac{Kb(h_2 - h_1)}{528\log(r_2/r_1)} \quad \text{for American Practical Units}$$

$$= \frac{2.73Kb(h_2 - h_1)}{\log(r_2/r_1)} \quad \text{for SI} \qquad \text{[Eq. III.2.1]}$$

where Q = pumping rate or well yield (in gpm, or m³/d), h_1, h_2 = static head measured from the aquifer bottom (in ft or m), r_1, r_2 = radial distance from the pumping well (in ft or m), b = thickness of the aquifer (in ft or m), and K = hydraulic conductivity of the aquifer (in gpd/ft² or m/d).

Many assumptions are made to derive the above equation. The following references or other groundwater hydrology books provide more detailed treatment of this subject:

1. *Groundwater and Wells*, 2nd ed., by F. G. Driscoll, Johnson Division, St. Paul, MN 55112, 1986.
2. *Applied Hydrogeology*, by C. W. Fetter, Jr., Charles E. Merrill Publishing, Columbus, OH, 1980.
3. *Groundwater Hydrology*, 2nd ed., by D. K. Todd, John Wiley & Sons, New York, 1980.
4. *Groundwater*, by R. A. Freeze and J. A. Cherry, Prentice Hall, Englewood Cliffs, NJ, 1979.

Hydraulic conductivity is often determined from aquifer tests (see Section III. 3 for details). Eq. III.2.1 can be easily modified to calculate hydraulic conductivity of a confined aquifer, if two steady-state drawdowns, flow rate, and aquifer thickness are available.

$$K = \frac{528Q\log(r_2/r_1)}{b(h_2 - h_1)} \quad \text{for American Practical Units}$$

$$= \frac{Q\log(r_2/r_1)}{2.73b(h_2 - h_1)} \quad \text{for SI} \qquad \text{[Eq. III.2.2]}$$

Another parameter, specific capacity, can also be used to estimate the hydraulic conductivity of an aquifer. Let us define the specific capacity as

$$\text{Specific capacity} = \frac{Q}{s_w} \qquad \text{[Eq. III.2.3]}$$

where Q = the well discharge rate (extraction rate), in gpm, and s_w = drawdown in the pumping well, in ft.

For example, if a well produced 50 gpm and the drawdown in the well is 5 ft, the specific capacity of this pumping well is 10 gpm/ft; it will produce 10 gpm for each foot of available drawdown. A rough estimate on transmissivity (in gpd/ft) can be obtained by multiplying the specific yield (in gpm/ft) by 2000 for confined aquifers and 1550 for unconfined aquifers.[7] The hydraulic conductivity (in gpd/ft^2) can then be determined by dividing the transmissivity with the aquifer thickness (in ft).

Example III.2.1A Steady-state drawdown from pumping a confined aquifer

A confined aquifer 30 ft (9.1 m) thick has a piezometric surface 80 ft (24.4 m) above the bottom confining layer. Groundwater is being extracted out from a 4-in (0.1 m) diameter fully penetrating well.

The pumping rate is 40 gpm (0.15 m^3/min). The aquifer is relatively sandy with a hydraulic conductivity of 200 gpd/ft^2. Steady-state drawdown of 5 ft (1.5 m) is observed in a monitoring well 10 ft (3.0 m) from the pumping well. Estimate

a. The drawdown 30 ft (9.1 m) away from the well
b. The drawdown in the pumping well

Solutions:

a. First let us determine h_1 (at r_1 = 10 ft):

$$h_1 = 80 - 5 = 75 \text{ ft} \quad (\text{or} = 24.4 - 1.5 = 22.9 \text{ m})$$

Use Eq. III.2.3:

$$40 = \frac{(200)(30)(h_2 - 75)}{528 \log(30/10)} \rightarrow h_2 = 76.7 \text{ ft}$$

or

$$(0.15)(1440) = \frac{2.73[(200)(0.0410)](9.1)(h_2 - 22.9)}{\log(9.1/3.0)} \rightarrow h_2 = 23.4 \text{ m}$$

So, drawdown at 30 ft (9.1 m) away = 80 – 76.7 = 3.3 ft (or = 24.4 – 23.4 = 1.0 m).

b. To determine the drawdown at the pumping well, set r at the well = well radius = (2/12) ft:

$$40 = \frac{(200)(30)(h_2 - 75)}{528 \log[(2/12)/10]} \rightarrow h_2 = 68.7 \text{ ft}$$

So, drawdown in the extraction well = 80 – 68.7 = 11.3 ft.

Discussion

1. In (a), the 0.041 is the conversion factor to convert the hydraulic conductivity from gpd/ft^2 to m/day. The factor was taken from Table III.1.A.
2. Calculations in (a) have demonstrated that the results would be the same by using two different systems of units.
3. The $(h_1 - h_2)$ term can be replaced by $(s_2 - s_1)$, where s_1 and s_2 are the drawdown at r_1 and r_2, respectively.
4. The same equation can also be used to determine the radius of influence, where drawdown is equal to zero. Discussions on this topic will be given in Chapter six.

Example III.2.1B *Estimate hydraulic conductivity of a confined aquifer from steady-state drawdown data*

Use the following information to estimate the hydraulic conductivity of a confined aquifer:

Aquifer thickness = 30.0 ft (9.1 m) thick
Well diameter = 4-in (0.1 m) diameter
Well perforation depth = fully penetrating
Groundwater extraction rate = 20 gpm
Steady-state drawdown = 2.0 ft observed in a monitoring well 5 ft from the pumping well

= 1.2 ft observed in a monitoring well 20 ft from the pumping well

Solutions:

Inserting the data into Eq. III.2.2, we obtain

$$K = \frac{528Q\log(r_2/r_1)}{b(h_2 - h_1)} = \frac{(528)(20)\log(20/5)}{(30)(2.0 - 1.2)} = 397 \text{ gpd/ft}^2$$

Discussion. The $(h_1 - h_2)$ term can be replaced by $(s_2 - s_1)$, where s_1 and s_2 are the drawdown at r_1 and r_2, respectively.

Example III.2.1C *Estimate hydraulic conductivity of a confined aquifer using specific capacity*

Use the drawdown data of the pumping well in Example III.2.1A to estimate the hydraulic conductivity of the aquifer:

Aquifer thickness = 30 ft
Pumping rate = 40 gpm
Steady-state drawdown in the well = 11.3 ft

Solutions:

a. First let us determine the specific capacity of this well. Use Eq. III.2.3:

$$\text{Specific capacity} = \frac{Q}{s_w} = \frac{40}{11.3} = 3.54 \text{ gpm/ft}$$

b. The transmissivity of the aquifer can be estimated as

$$T = (3.54)(2000) = 7080 \text{ gpd/ft}$$

c. The hydraulic conductivity of the aquifer can be estimated as

$$K = T/b = 7080/30 = 236 \text{ gpd/ft}^2$$

Discussion. The calculated hydraulic conductivity, 236 gpd/ft^2, from this exercise is not far from the value specified in Example III.2.1A, 200 gpd/ft^2.

III.2.2 Steady-state flow in an unconfined aquifer

The equation describing steady-state flow of an unconfined aquifer (water-table aquifer) from a fully penetrating well may be written as follows:

$$Q = \frac{K(h_2^2 - h_1^2)}{1055 \log(r_2/r_1)} \quad \text{for American Practical Units}$$

$$= \frac{1.366K(h_2^2 - h_1^2)}{\log(r_2/r_1)} \quad \text{for SI} \qquad \text{[Eq. III.2.4]}$$

All the terms are as defined for Eq. III.2.1.

Eq. III.2.4 can be easily modified to calculate the hydraulic conductivity of an unconfined aquifer if data of two steady-state drawdowns and flow rate are available.

$$K = \frac{1055Q \log(r_2/r_1)}{(h_2^2 - h_1^2)} \quad \text{for American Practical Units}$$

$$= \frac{Q \log(r_2/r_1)}{1.366(h_2^2 - h_1^2)} \quad \text{for SI} \qquad \text{[Eq. III.2.5]}$$

The specific capacity, defined by Eq. III.2.3, can also be used to estimate the hydraulic conductivity of an unconfined aquifer.

Example III.2.2A *Steady-state drawdown from pumping an unconfined aquifer*

A water-table aquifer is 80 ft (24.4 m) thick. Groundwater is being extracted out from a 4-in (0.1 m) diameter fully penetrating well.

The pumping rate is 40 gpm (0.15 m³/min). The aquifer is relatively sandy with a hydraulic conductivity of 200 gpd/ft². Steady-state drawdown of 5 ft (1.5 m) is observed in a monitoring well 10 ft (3.0 m) from the pumping well. Estimate

a. The drawdown 30 ft (9.1 m) away from the well
b. The drawdown in the pumping well

Solutions:

a. First let us determine h_1 (at r_1 = 10 ft):

$$h_1 = 80 - 5 = 75 \text{ ft} \quad (\text{or} = 24.4 - 1.5 = 22.9 \text{ m})$$

Use Eq. III.2.1:

$$40 = \frac{(200)(h_2^2 - 75^2)}{1055 \log(30/10)} \rightarrow h_2 = 75.7 \text{ ft}$$

or

$$(0.15)(1440) = \frac{1.366[(200)(0.0410)](h_2^2 - 22.9^2)}{\log(9.1/3.0)} \rightarrow h_2 = 23.1 \text{ m}$$

So, drawdown at 30 ft (9.1 m) away = 80 – 75.7 = 4.3 ft (or = 24.4 – 23.1 = 1.3 m)

b. To determine the drawdown at the pumping well, set r at the well = well radius = (2/12) ft

$$40 = \frac{(200)(h_2^2 - 75^2)}{1055 \log[(2/12)/10]} \rightarrow h_2 = 72.5 \text{ ft}$$

So, drawdown in the extraction well = 80 – 72.5 = 7.5 ft

Discussion.

1. In the equation for confined aquifers, the $(h_1 - h_2)$ term can be replaced by $(s_2 - s_1)$, where s_1 and s_2 are the drawdown at r_1 and r_2, respectively. However, no analogy can be made here for unconfined aquifers, that is, $(h_2^2 - h_1^2)$ can not be replaced by $(s_1^2 - s_2^2)$.

2. The same equation can also be used to determine the radius of influence, where drawdown is equal to zero. More discussions on this topic will be given in Chapter six.

Example III.2.2B *Estimate hydraulic conductivity of an unconfined aquifer from steady-state drawdown data*

Use the following information to estimate the hydraulic conductivity of an unconfined aquifer:

Aquifer thickness = 30.0 ft (9.1 m) thick
Well diameter = 4-in (0.1 m) diameter
Well perforation depth = fully penetrating
Groundwater extraction rate = 20 gpm
Steady-state drawdown = 2.0 ft observed in a monitoring well 5 ft from the pumping well

= 1.2 ft observed in a monitoring well 20 ft from the pumping well

Solutions:

First we need to determine h_1 and h_2:

$$h_1 = 30.0 - 2.0 = 28.0 \text{ ft}$$

$$h_2 = 30.0 - 1.2 = 28.8 \text{ ft}$$

Inserting the data into Eq. III.2.5, we obtain:

$$K = \frac{(1055)(20)\log(20/5)}{(28.8^2 - 28^2)} = 280 \text{ gpd/ft}^2$$

Discussion

1. In the equation for confined aquifers, the $(h_1 - h_2)$ term can be replaced by $(s_2 - s_1)$, where s_1 and s_2 are the drawdown at r_1 and r_2, respectively. However, no analogy can be made here, that is, $(h_2^2 - h_1^2)$ cannot be replaced by $(s_1^2 - s_2^2)$.
2. Drawdown and flow rate data in Examples III.2.1B and III.2.2B (one for a confined aquifer and the other for an unconfined aquifer) are the same; however, the calculated hydraulic conductivity values are different. In these examples, the hydraulic conductivity of the unconfined aquifer is lower, but it delivers the same flow rate with the same drawdown because the unconfined aquifer has a larger storage coefficient. Refer to Section III.1.4 for discussions of the storage coefficient.

Example III.2.2C Estimate hydraulic conductivity of an unconfined aquifer using specific capacity

Use the pumping and drawdown data in Example III.2.2A to estimate the hydraulic conductivity of the aquifer:

Aquifer thickness = 80 ft
Pumping rate = 40 gpm
Steady-state drawdown in the well = 7.5 ft

Solutions:

a. First let us determine the specific capacity of this well. Use Eq. III.2.3,

$$\text{Specific capacity} = \frac{Q}{s_w} = \frac{40}{7.5} = 5.3 \text{ gpm/ft}$$

b. The transmissivity of the aquifer can be estimated as

$$T = (5.3)(1550) = 8220 \text{ gpd/ft}$$

c. The hydraulic conductivity of the aquifer can be estimated as

$$K = T/b = 8220/80 = 103 \text{ gpd/ft}^2$$

Discussion. The calculated hydraulic conductivity, 103 gpd/ft^2, from this exercise has the same order of magnitude as the value specified in Example III.2.2A, 200 gpd/ft^2.

III.3 Aquifer test

In Section III.2, methods using the steady-state drawdown data (Eqs. III.2.2 and III.2.5) were described to estimate the hydraulic conductivity of aquifers. For a groundwater remediation project, it is often required to have a good estimate of the hydraulic conductivity before the full-scale groundwater extraction. Grain-size analysis of aquifer materials and bench-scale testing on core samples can provide some limited information. For more accurate estimates, aquifer tests are often conducted.

Pumping tests and slug tests are two common types of aquifer tests. In a typical pumping test, groundwater is extracted from a pumping well at a constant rate. (Other pumping schemes are also feasible, but not as popular.) The time-dependent drawdowns (or recovery) in the pumping well and/or in a few monitoring wells are recorded. The data are then analyzed to determine the hydraulic conductivity and storativity. The pumping test is recommended because it provides information on subsurface hydrogeology over a large area (the area affected by the pumping) and gives a realistic estimate of the pumping rate for the full-scale groundwater extraction. Many remediation systems have been incorrectly designed and installed for a flow rate much higher than the extraction wells could yield for lack of accurate aquifer information. In addition, analysis of groundwater extracted during a pump test will give the engineers a more realistic estimation of the contaminant concentrations for treatment system design than those just based on the data from sampling of monitoring wells. The disadvantage of a pumping test is mainly the cost of conducting the test, data analysis, and treatment and disposal of the extracted water, which is usually contaminated.

A cheaper alternative to a pumping test is a slug test in which a slug of known volume is injecting into a well. The rate at which the water level falls is collected and analyzed. The disadvantages of a slug test are (1) it provides only the hydrological information related to the vicinity of the well and (2) it provides no additional information for estimates of the contaminant concentration as the full-scale remediation program starts. No further discussions on slug tests will be given here.

The flow in the aquifer during a pumping test is considered to be under unsteady-state conditions. Three common methods are used to analyze the unsteady-state data: (1) Theis curve matching, (2) the Jacob straight-line method, and (3) the distance-drawdown method.

III.3.1 Theis Method

The drawdown for confined aquifers under unsteady-state pumping was first solved by C.V. Theis as

$$s = \frac{114.6Q}{T}\left[-0.5772 - \ln(u) + u - \frac{u^2}{2\cdot 2!} + \frac{u^3}{3\cdot 3!} - \frac{u^4}{4\cdot 4!} + \ldots\right]$$

in American Practical Units [Eq.III.3.1]

$$= \frac{Q}{4\pi T}\left[-0.5772 - \ln(u) + u - \frac{u^2}{2\cdot 2!} + \frac{u^3}{3\cdot 3!} - \frac{u^4}{4\cdot 4!} + \ldots\right] \text{ in SI}$$

where the argument u is dimensionless and given as

$$u = \frac{1.87 r^2 S}{Tt} \text{ in American Practical Units}$$

[Eq. III.3.2]

$$= \frac{r^2 S}{4Tt} \text{ in SI}$$

and s = drawdown at time t (in ft or m), Q = constant pumping rate (in gpm or m^3/d), r = radial distance from the pumping well to the observation well (in ft or m), S = aquifer storativity (dimensionless), T = aquifer transmissivity (in gpd/ft or m^2/d), and t = time since pumping started (in days).

The infinite series term in Eq. III.3.1 (the terms inside the square bracket) is often called the well function and designated as $W(u)$. Tabulated values of $W(u)$ as a function of u can be found in most of the groundwater hydrology books. (The well function tables have become obsolete because of the convenience of hand calculators and personal computers.) A type-curve approach is often developed to match the time and drawdown data to the curve of $W(u)$ vs. $1/u$. From the match points the transmissivity and storativity can be determined. There are several computer programs available on the market for Theis curve matching. This subsection will provide one example of using the Theis equation, but no examples for curve matching will be given.

Example III.3.1 *Estimate unsteady-state drawdown of a confined aquifer using the Theis equation*

A pumping well is installed in a confined aquifer. Use the following information to estimate the drawdown at a distance 20 feet away from the well after one day of pumping:

Aquifer thickness = 30.0 ft
Groundwater extraction rate = 20 gpm
Aquifer hydraulic conductivity = 400 gpd/ft^2
Aquifer storativity = 0.005

Solutions:

a. $T = Kb$ = (400)(30) = 12,000 gpd/ft.

b. Inserting the data into Eq. III.3.2, we obtain

$$u = \frac{1.87\,r^2S}{Tt} = \frac{1.87(20\,\text{ft})^2(0.005)}{(12{,}000\,\text{gpd/ft})(1\text{ day})} = 3.12\times10^{-4}$$

c. Substitute the value of u in the well function to obtain its value:

$$W(u) = \Big[-0.5772 - \ln(3.12\times10^{-4}) + (3.12\times10^{-4}) - \frac{(3.12\times10^{-4})^2}{2\cdot 2!}$$

$$+\frac{(3.12\times10^{-4})^3}{3\cdot 3!} - \frac{(3.12\times10^{-4})^4}{4\cdot 4!} + \ldots\Big] = 7.50$$

d. The drawdown can then be determined from Eq. III.3.1:

$$s = (114.6)(20)(7.50)/(12{,}000) = 1.43\text{ ft}$$

Discussion. For the small u values, the third and later terms in the well function can be truncated without causing a significant error.

III.3.2 Cooper–Jacob straight-line method

As shown in the last example, the higher terms in the well function become negligible for small u values. Cooper and Jacob in 1946 pointed out that, for small u values, the Theis equation can be modified to the following form without significant errors:

$$s = \frac{264\,Q}{T}\log\left[\frac{0.3\,Tt}{r^2 S}\right] \text{ in American Practical Units}$$

$$= \frac{0.183\,Q}{T}\log\left[\frac{2.25\,Tt}{r^2 S}\right] \text{ in SI} \qquad \text{[Eq. III.3.3]}$$

where the symbols represent the same terms as in Eq. III.3.1.

As shown in Eq. III.3.2, the value of u becomes small as t increases and r decreases. So Eq. III.3.3 is valid after sufficient pumping time and at a short distance from the well ($u < 0.05$). It can be seen from Eq. III.3.3, at any specific location (r = constant), s varies linearly with log[(constant)t]. The Jacob's straight line method is to plot drawdown vs. pumping time data from a pumping test on semilog paper; most of the data should fall on a straight line. From the plot, the slope, Δs (the change in drawdown per one log cycle of time), and the intercept, t_o, of the straight line at zero drawdown can be derived. The following relationships can then be used to determine the transmissivity and storativity of the aquifer:

$$T = \frac{264\,Q}{\Delta s} \text{ in American Practical Units } = \frac{0.183\,Q}{\Delta s} \text{ in SI} \qquad \text{[Eq.III.3.4]}$$

$$S = \frac{0.3\,Tt_o}{r^2} \text{ in American Practical Units } = \frac{2.25\,Tt_o}{r^2} \text{ in SI} \qquad \text{[Eq.III.3.5]}$$

where Δs is in ft or in m, t_o in days, and the other symbols represent the same terms as in Eq. III.3.1.

Example III.3.2 Analysis of pumping test data using Cooper-Jacob's straight-line method

A pumping test (Q = 120 gpm) was conducted on a confined aquifer (aquifer thickness = 30.0 ft). The time-drawdown data at a distance 150 ft away from the well were collected and are shown in the following table.

Use the Cooper–Jacob straight-line method to determine the hydraulic conductivity and storativity of the aquifer.

Time since pumping started (min)	Drawdown, s (ft)
7	0.15
20	0.45
80	0.90
200	1.16

Solutions:

a. The data are first plotted on a semilog scale.

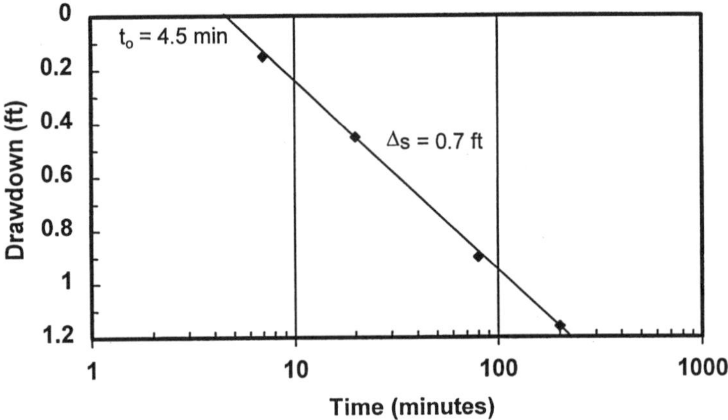

Figure E.III.3.2 Cooper–Jacob straight line method for pumping data analysis.

From the plot, we find $\Delta s = 0.7$ ft.

b. Use Eq. III.3.4

$$T = \frac{264Q}{\Delta s} = \frac{(264)(50)}{0.7} = 18{,}860\,\text{gpd/ft}$$

c. Hydraulic conductivity can then be found as

$$K = T/b = (18{,}860)/(30) = 629\ \text{gpd/ft}^2$$

d. From the plot, we find the intercept, $t_o = 4.5$ min $= 3.1 \times 10^{-3}$ day. Use Eq. III.3.5 to find the storativity:

$$S = \frac{0.3Tt_o}{r^2} = \frac{(0.3)(18{,}860)(0.0031)}{(150)^2} = 0.00078$$

Discussion. At $t = 7$ minutes (0.00486 day) and $r = 150$ ft, u is equal to

$$u = \frac{1.87\,r^2 S}{Tt} = \frac{1.87(150\,\text{ft})^2(0.00078)}{(18{,}860\,\text{gpd/ft})(0.00486\,\text{day})} = 0.36$$

At t = 60 min, u will be smaller than 0.05.

III.3.3 Distance-drawdown method

It can be seen from Eq. III.3.3, at any specific location (r = constant), s varies linearly with log[(constant)/r^2]. Based on this relationship and simultaneous drawdown measurements in at least three observation wells each at a different distance from the pumping well, a semilog distance-drawdown graph can be constructed. From the plot, the slope, Δs (the change in drawdown per one log cycle of distance), and the intercept, r_o, of the straight line at zero drawdown can be derived. The following relationships can then be used to determine the transmissivity and storativity of the aquifer:

$$T = \frac{528\,Q}{\Delta s} \text{ in American Practical Units } = \frac{0.366\,Q}{\Delta s} \text{ in SI} \quad \text{[Eq.III.3.6]}$$

$$S = \frac{0.3\,Tt}{r_o^2} \text{ in American Practical Units } = \frac{2.25\,Tt}{r_o^2} \text{ in SI} \quad \text{[Eq.III.3.7]}$$

where Δs is in ft or in m, r_o is in ft or in m, and the other symbols represent the same terms as in Eq. III.3.1.

The three methods described here for analysis of pumping test data are mainly for confined aquifers. A well pumping from an unconfined aquifer is more complicated. The extracted water comes from two mechanisms: (1) water from the elastic storage due to the decline in pressure, as in the case of the confined aquifer, and (2) water from drainage of the declining water table. There are three distinct phases of time-drawdown relations in unconfined aquifers. However, as time progresses, the rate of drawdown decreases and flow becomes essentially horizontal (when the effects of gravity drainage become much smaller). The time-drawdown data can then be analyzed using the three methods described above.[2] A more practical approach is to ensure that the duration of the pumping test exceeds the suggested guidelines in Table III.3.A.[5] As shown in the table, the suggested pumping duration increases with the tightness of the aquifer. A minimum of 7 days pumping is suggested for silty or clayey aquifers.

Example III.3.3 Analysis of pumping test data using the distance-drawdown method

A pumping test (Q = 120 gpm) was conducted on a confined aquifer (aquifer thickness = 30.0 ft). The distance-drawdown (at t = 90 minutes) were collected from three monitoring wells and shown in the following table.

Distance from the pumping well (ft)	Drawdown, s (ft)
50	1.55
150	0.90
300	0.50

Use the distance-drawdown method to determine the hydraulic conductivity and storativity of the aquifer.

Solutions:

a. The data are first plotted on a semilog scale.

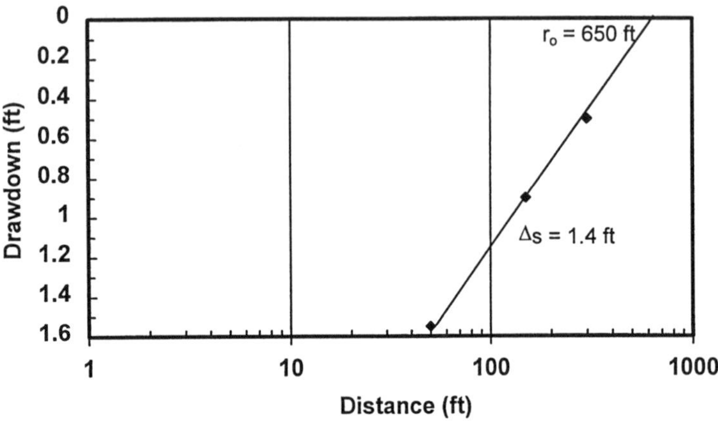

Figure E.III.3.3 Distance-drawdown method for pumping data analysis.

From the plot, we find Δs = 1.4 ft.

b. Use Eq. III.3.6 to find the transmissivity:

$$T = \frac{528Q}{\Delta s} = \frac{(528)(50)}{1.4} = 18{,}860\,\text{gpd/ft}$$

c. Hydraulic conductivity can then be found as

$$K = T/b = (18{,}860)/(30) = 629\ \text{gpd/ft}^2$$

d. From the plot, we find the intercept, r_o = 650 ft.

e. Use Eq. III.3.7 and t = 90 min = 0.0625 day to find the storativity:

$$S = \frac{0.3\,Tt}{r_o^2} = \frac{(0.3)(18{,}860)(0.0625)}{(650)^2} = 0.00084$$

Discussion

1. As expected, the slope of the straight line in the distance drawdown is twice that in the Cooper–Jacob straight line plot (for the same hydraulic conductivity and pumping rate).
2. At t = 90 minutes (0.0625 day) and r = 300 ft, u is equal to

$$u = \frac{1.87\,r^2 S}{Tt} = \frac{1.87(300\,\text{ft})^2(0.00074)}{(18{,}860\,\text{gpd/ft})(0.0625\,\text{day})} = 0.11$$

At $r < 204$ ft, u will be smaller than 0.05.

Table III.3.A Suggested Guidelines for Pumping Tests of Unconfined Aquifers

Predominant aquifer material	Minimum pumping time (hours)
Medium sand and coarser materials	4
Fine sand	30
Silt and clay	170

From Todd, D. K., *Groundwater Hydrology*, 2nd ed., John Wiley & Sons, New York, 1980. With permission.

III.4 Migration velocity of the dissolved plume

As VOC spills enter into the subsurface, the materials may move downward as free product or be dissolved into the infiltrating water and then move downward by gravity. This liquid may travel deep enough to come in contact with the underlying aquifer and form a dissolved plume in the aquifer. This section will discuss the transport of contaminant in the dissolved plume, which is relatively simpler than the transport in the vadose zone. This discussion applies not only to VOCs, but to other contaminants such as heavy metals. Transport in the vadose zone will be discussed later in Section III.5.

III.4.1 The advection–dispersion equation

Design and selection of optimal remediation schemes, such as the number and locations of extraction wells, often require predicting the contaminant distribution within the subsurface over the specified period of time. These predictions are then used to evaluate different remediation scenarios. To make such predictions we need to couple the equation describing the flow with the concept of mass balance. More discussions on the mass balance concept can be found in Chapter four.

To fully describe a contaminant migration, the one-dimensional form of the advection–dispersion equation can be expressed as

$$\frac{\partial C}{\partial t} = D\frac{\partial^2 C}{\partial x^2} - v\frac{\partial C}{\partial x} \pm RXNs \quad \text{[Eq. III.4.1]}$$

where C is the contaminant concentration, D is the dispersion coefficient, v is the velocity of the flow, t is time, and *RXNs* represents the reactions. Eq. III.4.1 is a general equation and it is applicable to describe the fate and transport of contaminants in the vadose zone or in the groundwater. The first term of Eq. III.4.1 describes the change in contaminant concentration in the water contained within the volume of the aquifer or the vadose zone. The first term on the right-hand side describes the net dispersive flux in and out of a volume of the aquifer or the vadose zone. The second term on the right-hand side describes the net advective flux of the contaminant, and the last term represents the amount of contaminant that may be added or lost to the water by some physical, chemical, or biological reactions. For plume migration in groundwater, v is the groundwater velocity that can be determined from Darcy's Law and the porosity of the aquifer (i.e., Eq. III.1.3).

III.4.2 Diffusivity and dispersion coefficient

The dispersion term in Eq. III.4.1 accounts for both the molecular diffusion and hydraulic dispersion. The molecular diffusion, strictly speaking, is due to the concentration gradient (i.e., the concentration difference). The compound tends to diffuse away from the higher concentration zone, and this can occur even when the water is not moving. The hydraulic dispersion here is mainly caused by flow in porous media. It results from (1) velocity variation within a pore, (2) different pore geometrics, (3) the divergence of flow lines around the soil grains present in porous media, and (4) the aquifer heterogeneity.[6]

The unit of the dispersion coefficient is (length)^2/(time). Field studies of the dispersion coefficient revealed that it varies with the groundwater velocity. They show that the dispersion coefficient is relatively constant at low velocities (where the molecular diffusion dominates), but increases linearly with velocity as the groundwater velocity increases (when the hydraulic dispersion dominates). The dispersion coefficient can be written as the sum of two terms: an effective molecular diffusion coefficient, D_d, and hydraulic dispersion coefficient, D_h:

$$D = D_d + D_h \quad \text{[Eq. III.4.2]}$$

The effective molecular diffusion coefficient can be obtained from the molecular diffusion coefficient, D_o, by

$$D_d = (\xi)(D_o) \quad \text{[Eq. III.4.3]}$$

where ξ is the tortuosity factor that accounts for the increased distance that the contaminant must travel to get around the soil grains. Typical ξ values are in the range of 0.6 to 0.7.[6]

The hydraulic dispersion coefficient is proportional to the velocity as

$$D_h = (\alpha)(\nu) \qquad \text{[Eq. III.4.4]}$$

where α is the dispersivity. The hydraulic dispersion coefficient is scale dependent; its value has been observed to increase with increasing transport distance. The longitudinal dispersivity values from field tracer tests and model calibration of contaminant plumes are found to be in the range of 10 to 100 m, which is much higher than those from column studies in the laboratory.

The molecular diffusion coefficient for contaminant in dilute aqueous solutions is very much smaller than in gases at atmospheric pressure, usually falling in the range of 0.5 to 2×10^{-5} cm^2/s at 25°C (compared with typical values of 0.05 to 0.5 cm^2/s for diffusion in the gaseous phase, as shown in Table II.3.C). Values of molecular diffusion coefficients of selected compounds are shown in Table III.4.A.

The diffusion coefficient of contaminant in water can be estimated by using the Wilke–Chang method.[4]

$$D_o = \frac{5.06 \times 10^{-7}\, T}{\mu_w V^{0.6}} \qquad \text{[Eq. III.4.5]}$$

where D_o = the diffusion coefficient, in cm^2/s, T = the temperature in K, μ_w = the viscosity of water, in *cp* (see Table III.1.B), and V = the molal volume of the solute at its normal boiling point, in cm^3/g mole.

Table III.4.A Values of Diffusion Coefficients of Selected Compounds in Water

Compound	Temperature (°C)	Diffusion coefficient (cm^2/s)
Acetone	25	1.16×10^{-5}
Acetonitrile	15	1.26×10^{-5}
Benzene	20	1.02×10^{-5}
Benzoic acid	25	1.00×10^{-5}
Butanol	15	0.77×10^{-5}
Ethylene glycol	25	1.16×10^{-5}
Propanol	15	0.87×10^{-5}

Adapted from Sherwood, T. K., Pigford, R. L., and Wilke, C. R., *Mass Transfer*, McGraw-Hill, New York, 1975. With permission.

Table III.4.B Additive-Volume Increments for Calculation of Molal Volumes

	Increment (cm^3/g mole)
Carbon	14.8
Hydrogen	3.7
Oxygen (excepted as noted below)	7.4
In methyl esters and ethers	9.1
In ethyl esters and ethers	9.9
In higher esters and ethers	11.0
In acids	12.0
Joined to S, P, and N	8.3
Nitrogen	
Doubly bonded	15.6
In primary amines	10.5
In secondary amines	12.0
Bromine	27
Chlorine	24.6
Ring	
Three–membered	–6.0
Four–membered	–8.5
Five–membered	–11.5
Six–membered	–15.0
Naphthalene	–30.0
Anthracite	–47.5

From Sherwood, T. K., Pigford, R. L., and Wilke, C. R., *Mass Transfer*, McGraw-Hill, New York, 1975. With permission.

The molal volume can be used from the method of LeBas by using Table III.4.B.

The diffusion coefficient can also be estimated from the diffusion coefficient of another compound of similar species and molecular weight by the following relationship:

$$\frac{D_1}{D_2} = \sqrt{\frac{MW_2}{MW_1}} \qquad \text{[Eq. III.4.6]}$$

As shown in Eq. III.4.6, the diffusion coefficient is inversely proportional to the square root of its molecular weight. The heavier the contaminant, the harder it is to diffuse through the water. Temperature also has an influence on the diffusion coefficient. From Eq. III.4.5, we can see the diffusion coefficient in water is proportional to the temperature and inversely proportional to the water viscosity. The water viscosity decreases with increasing temperature, and, consequently, the diffusion coefficient increases with temperature and the following relationship applies:

$$\frac{D_o @T_1}{D_o @T_2} = \left(\frac{T_1}{T_2}\right)\left(\frac{\mu_w @T_2}{\mu_w @T_1}\right) \qquad \text{[Eq. III.4.7]}$$

Example III.4.2A *Estimate the diffusion coefficient using the LeBas method*

Estimate the diffusion coefficient of toluene in dilute aqueous solution at 20°C using the LeBas method.

Solution:

a. The formula of toluene is $C_6H_5CH_3$. It consists of a benzene ring (six carbon member) and a methyl group. Viscosity of water at 25°C = 0.89 cp (from Table III.1.B).

$$T = 273 + 20 = 293K$$

Molal volume is determined from the sum of the volume increment (Table III.4.B):

$$C = (14.8)(7) = 103.6$$

$$H = (3.7)(8) = 29.6$$

$$\text{Six-membered ring} = -15.0$$

So,

$$V = 103.6 + 29.6 - 15.0 = 118.2 \text{ cm}^3/\text{g mole}$$

b. Use Eq. III.4.7 to find the diffusion coefficient:

$$D_o = \frac{5.06 \times 10^{-7}(293)}{(0.89)(118.2)^{0.6}} = 0.95 \times 10^{-5} \text{ cm}^2/\text{s}$$

Example III.4.2B *Estimate the diffusion coefficient at different temperatures*

The diffusion coefficient of benzene in dilute aqueous solution at 20°C is 1.02×10^{-5} cm²/s (Table III.4.A). Use this reported value to estimate

a. The diffusion coefficient of toluene in dilute aqueous solution at 20°C
b. The diffusion coefficient of benzene in dilute aqueous solution at 25°C.

Solution:

a. The molecular weight of toluene ($C_6H_5CH_3$) is 92, and the molecular weight of benzene (C_6H_6) is 78. Use Eq. III.4.6 to find the diffusion coefficient:

$$\frac{D_1}{D_2} = \frac{(1.02 \times 10^{-5})}{D_2} = \sqrt{\frac{92}{78}}$$

So, the diffusion coefficient of toluene at 20°C = 0.94 × 10^{-5} cm^2/s.

b. Viscosity of water at 20°C = 1.002 cp (from Table III.1.B). Viscosity of water at 25°C = 0.89 cp (from Table III.1.B)
Use Eq. III.4.5 to find the diffusion coefficient:

$$\frac{(1.02 \times 10^{-5})}{D_o @ 298\,K} = \left(\frac{293}{298}\right)\left(\frac{0.89}{1.002}\right)$$

So, the diffusion coefficient of benzene at 25°C = 1.17 × 10^{-5} cm^2/s.

Discussion

1. The diffusion coefficient of toluene estimated from that of benzene is 0.94 × 10^{-5} cm^2/s, which is essentially the same as that from the LeBas method, 0.94 × 10^{-5} cm^2/s (Example III.4.2A).
2. The diffusion coefficient of benzene at 25°C is about 15% higher than that at 20°C.

Example III.4.2C *Relative importance of the molecular diffusion and hydraulic dispersion*

Benzene from leaky USTs at a site leaked into the underlying aquifer. The hydraulic conductivity of the aquifer is 500 gpd/ft^2 and it has a porosity of 0.4. The groundwater temperature is 20°C. The dispersivity is found to be 2 m. Estimate the relative importance between the hydraulic dispersion and the molecular diffusion for the dispersion of the benzene plume in the following two cases:

1. The hydraulic gradient = 0.01
2. The hydraulic gradient = 0.0005

Solution:

a. The hydraulic conductivity of the aquifer = 500 gpd/ft^2 = (500)(4.73*E* – 5) = 0.024 cm/s (Use the conversion factor from Table III.1.A). Use Eqs. III.1.1 and III.1.2 to find the groundwater velocity (for gradient = 0.01).

$$v_s = \frac{(0.024)(0.01)}{0.4} = 6 \times 10^{-4}\ \text{cm/s}$$

The molecular diffusion coefficient of benzene (at 20°C) = 1.02×10^{-5} cm^2/s (Table III.4.A). From Eq. III.4.2,

$$D = D_d + D_h \qquad \text{[Eq. III.4.2]}$$

But, the effective molecular diffusion coefficient can be obtained as (Eq. III.4.3) by assuming $\xi = 0.65$:

$$D_d = \xi\,(D_o) = (0.65)(1.02 \times 10^{-5}) = 0.66 \times 10^{-5}\ \text{cm}^2/\text{s}$$

The hydraulic dispersion coefficient can then be determined as (Eq. III.4.4) by assuming $\alpha = 2$ m:

$$D_h = \alpha(v) = (200\ \text{cm})(6 \times 10^{-4}\ \text{cm/s}) = 12{,}000 \times 10^{-5}\ \text{cm}^2/\text{s}$$

The hydraulic dispersion coefficient is much larger than the diffusion coefficient. Therefore, the hydraulic dispersion will be the dominant mechanism for contaminant dispersion.

b. For a smaller gradient, the groundwater will move more slowly, and the dispersion coefficient will be proportionally smaller. The effective molecular diffusion coefficient will be the same as 0.66×10^{-5} cm^2/s. Use Eqs. III.1.1 and III.1.2 to find the groundwater velocity (for gradient = 0.0005):

$$v_s = \frac{(0.024)(0.0005)}{0.4} = 3.0 \times 10^{-5}\ \text{cm/s}$$

The hydraulic dispersion coefficient can then be determined as (Eq. III.4.4):

$$D_h = \alpha(v) = (200\ \text{cm})(3.0 \times 10^{-5}\ \text{cm/s}) = 600 \times 10^{-5}\ \text{cm}^2/\text{s}$$

The hydraulic dispersion coefficient is still much larger than the diffusion coefficient at this flat gradient (0.0005).

Discussion. In the second case, the groundwater movement is very slow at 3.0×10^{-5} cm/s (or 31 ft/yr); the hydraulic dispersion is still the dominant mechanism (for dispersivity = 2 m). The diffusion coefficient will become more important only if the flow rate and/or the dispersivity is smaller. Nonetheless, the molecular diffusion accounts for a common phenomenon that the plume usually extends slightly upstream of the discharge point.

III.4.3 Retardation factor for migration in groundwater

Physical, chemical, and biological processes in the subsurface that can affect the fate and transport of contaminants include (1) biotic degradation, (2) abiotic degradation, (3) dissolution, (4) ionization, (5) volatilization, and (6) adsorption. For transport of dissolved plume in groundwater, adsorption of contaminants is probably the most important and most studied mechanism for removal of contaminants from the groundwater. If adsorption is the primary removal mechanism in the subsurface, the reaction term in Eq. III.4.1 can then be written as $(\rho_b/\phi)\partial X/\partial t$, where ρ_b is the dry bulk density of soil (or the aquifer matrix), ϕ is the porosity, t is time, and X is the contaminant concentration in soil.

When the contaminant concentration is low, a linear adsorption isotherm is usually valid (see Section II.3.3 for more discussions on the adsorption isotherms). Assume a linear adsorption isotherm (e.g., $X = K_pC$), thus

$$\frac{\partial X}{\partial C} = K_p \qquad \text{[Eq. III.4.8]}$$

The following relationship can then be derived:

$$\frac{\partial X}{\partial t} = \left(\frac{\partial X}{\partial C}\right)\left(\frac{\partial C}{\partial t}\right) = K_p\frac{\partial C}{\partial t} \qquad \text{[Eq. III.4.9]}$$

Substitute Eq. III.4.9 into Eq. III.4.1 and rearrange the equation

$$\frac{\partial C}{\partial t} + \left(\frac{\rho_b}{\phi}\right)K_p\frac{\partial C}{\partial t} = \left(1+\frac{\rho_b K_p}{\phi}\right)\frac{\partial C}{\partial t} = D\frac{\partial^2 C}{\partial x^2} - \mathrm{v}\frac{\partial C}{\partial x} \qquad \text{[Eq. III.4.10]}$$

Dividing both sides by $(1 + \rho_b K_p/\phi)$, Eq. III.4.10 can be simplified into the following form:

$$\frac{\partial C}{\partial t} = \frac{D}{R}\frac{\partial^2 C}{\partial x^2} - \frac{\mathrm{v}}{R}\frac{\partial C}{\partial x} \qquad \text{[Eq. III.4.11]}$$

where

$$R = 1 + \frac{\rho_b K_p}{\phi} \qquad \text{[Eq. III.4.12]}$$

The parameter, R, is often called the retardation factor (dimensionless) and has a value ≥ 1. Eq. III.4.12 is essentially the same as Eq. III.4.1 except that the reaction term in Eq. III.4.1 is taken care of by R in Eq. III.4.12. The retardation factor reduces the impact of dispersion and migration velocity by a factor of R. (All of the mathematical solutions that are used to solve the transport of inert tracers can be used for the transport of the contaminants if the groundwater velocity and the dispersion coefficient are divided by the retardation factor.) From the definition of R, we can tell that R is a function of ρ_b, ϕ, and K_p. For a given aquifer, ρ_b and ϕ would be the same for different contaminants. Consequently, the greater the partition coefficient, the greater the retardation factor.

Example III.4.3 Determination of the retardation factor

The groundwater underneath a landfill is contaminated by landfill leachates containing benzene, 1,2-dichloroethane (DCA), and pyrene.

Estimate the retardation factor using the following data from the site assessment:

Aquifer porosity = 0.40
Bulk density of aquifer materials = 1.8 g/cm^3
Fraction of organic carbon of aquifer materials = 0.015
$K_{oc} = 0.63\ K_{ow}$

Solution:

a. From Table II.3.C,

$$\text{Log}(K_{ow}) = 2.13 \text{ for benzene} \rightarrow K_{ow} = 135$$

$$\text{Log}(K_{ow}) = 1.53 \text{ for 1,2-DCA} \rightarrow K_{ow} = 34$$

$$\text{Log}(K_{ow}) = 4.88 \text{ for pyrene} \rightarrow K_{ow} = 75{,}900$$

b. Using the given relationship, $K_{oc} = 0.63K_{ow}$, we obtain:

$$K_{oc} = (0.63)(135) = 85 \text{ (for benzene)}$$

$$K_{oc} = (0.63)(34) = 22 \text{ (for 1,2-DCA)}$$

$$K_{oc} = (0.63)(75{,}900) = 47{,}800 \text{ (for pyrene)}$$

c. Using Eq. II.3.12, $K_p = f_{oc}K_{oc}$, and $f_{oc} = 0.015$, we obtain:

$$K_p = (0.015)(85) = 1.275 \text{ (for benzene)}$$

$$K_p = (0.015)(22) = 0.32 \text{ (for 1,2-DCA)}$$

$$K_p = (0.015)(47{,}800) = 717 \text{ (for pyrene)}$$

d. Use Eq. III.4.12 to find the retardation factor

$$R = 1 + \frac{\rho_b K_p}{\phi} = 1 + \frac{(1.8)(1.275)}{0.4} = 6.74 \quad \text{for benzene}$$

$$R = 1 + \frac{\rho_b K_p}{\phi} = 1 + \frac{(1.8)(0.32)}{0.4} = 2.44 \quad \text{for 1,2-DCA}$$

$$R = 1 + \frac{\rho_b K_p}{\phi} = 1 + \frac{(1.8)(717)}{0.4} = 3227 \quad \text{for pyrene}$$

Discussion. Pyrene is very hydrophobic and its retardation factor is much higher than that of benzene or 1,2-DCA.

III.4.4 Migration of the dissolved plume

The retardation factor relates the plume migration velocity to the groundwater seepage velocity as

$$R = \frac{V_S}{V_p} \quad \text{or} \quad V_p = \frac{V_S}{R} \qquad \text{[Eq. III.4.13]}$$

where V_s is the groundwater seepage velocity and V_p is the velocity of the dissolved plume. When the value of R is equal to unity (for inert compounds), the compound will move at the same speed as the groundwater flow without any "retardation"; when $R = 2$, for example, the contaminant will move at half of the groundwater flow velocity.

Example III.4.4A Migration speed of the dissolved plume in groundwater

The groundwater underneath a landfill is contaminated by landfill leachates containing benzene, 1,2-DCA, and pyrene. A recent groundwater monitoring in September 1997 indicated that 1,2-DCA and benzene have traveled 250 and 20 m down gradient, respectively, while no pyrene compounds were detected in the down-gradient well.

Estimate the time when the leachates first entered the aquifer. The following data were obtained during the site assessment phase:

Aquifer porosity = 0.40
Aquifer hydraulic conductivity = 30 m/day

Groundwater gradient = 0.01
Bulk density of aquifer materials = 1.8 g/cm^3
Fraction of organic carbon of aquifer materials = 0.015
K_{ow} = 153,000 for pyrene
K_{oc} = 0.63 K_{ow}

Briefly discuss your results and list possible factors that may cause your estimate to differ from the true value.

Solution:

a. Use Eq. III.1.1 to find the Darcy velocity:

$$v = ki = (30)(0.01) = 0.3 \text{ m/d}$$

b. Use Eq. III.1.3 to find the groundwater velocity (or the seepage velocity, the interstitial velocity)

$$v_s = v/\phi = (0.3)/(0.4) = 0.75 \text{ m/d}$$

c. Use Eq. III.4.13 and the values of *R* from Example III.4.3 to determine the migration speeds of the plumes:

$$v_p = (0.75)/(6.74) = 0.111 \text{ m/d} = 40.6 \text{ m/yr (for benzene)}$$

$$v_p = (0.75)/(2.44) = 0.307 \text{ m/d} = 112 \text{ m/yr (for 1,2-DCA)}$$

$$v_p = (0.75)/(6508) = 0.000115 \text{ m/d} = 0.04 \text{ m/yr (for pyrene)}$$

d. The time for 1,2-DCA to travel 250 m can be found as:

$$t = (\text{distance})/(\text{migration speed}) = (250 \text{ m})/(112 \text{ m/yr}) = 2.23 \text{ yr} = 2 \text{ years and 3 months}$$

So, 1,2-DCA entered the groundwater in June of 1995.

e. The time for benzene to travel 50 m can be found as

$$t = (50 \text{ m})/(40.6 \text{ m/yr}) = 1.23 \text{ yr} = 1 \text{ year and 3 months}$$

So benzene entered the groundwater in June of 1996.

Discussion

1. The estimates are the time when the benzene and 1,2-DCA entered the aquifer. The information given is not sufficient to estimate the time the leachates leaked through the landfill liner.
2. The retardation of 1,2-DCA is smaller; therefore, its migration speed in the vadose zone would be higher. This also explains the fact that 1,2-DCA entered the aquifer earlier than benzene.

3. The migration of pyrene is extremely small, 0.04 m/yr; therefore, it was not detected in the downstream monitoring wells. Most, if not all, of the pyrene compounds will be adsorbed onto the soil in the vadose zone.
4. The estimates are crude, because many factors may affect the accuracy of the estimates. Factors include uncertainty of the hydraulic conductivity, porosity, groundwater gradient, K_{ow}, f_{oc}, etc. Neighborhood pumping will affect the natural groundwater gradient and, consequently, the migration of the plume. Other subsurface reactions such as oxidation and biodegradation may also have large impacts on the fate and transport of these contaminants.

Example III.4.4B Migration speed of the dissolved plume in groundwater

Results of a recent quarterly groundwater monitoring (July of 1997) at a contaminated site indicate that the edge of the dissolved TCE plume had advanced 200 m in the past 5 years. The groundwater gradient was determined to be 0.02 from this round of monitoring. Using a value of 4.0 for the retardation factor and aquifer porosity of 0.35, what would be your estimate for the hydraulic conductivity of the aquifer ? Also, because of the drought, an adjacent facility (down gradient from the site) pumped out a great amount of groundwater in 1995. How will this affect your estimate?

Solution:

a. The migration speed of the plume, v_p:

$$(\text{distance})/(\text{time}) = (200)/5 = 40 \text{ m/yr}$$

b. Use Eq. III.4.13 and the value of R to find the groundwater velocity, v_s:

$$v_p = v_s/R = 40 = v_s/4 \rightarrow v_s = 160 \text{ m/yr}$$

c. Use Eq. III.1.3 to find the Darcy velocity, v:

$$v_s = v/\phi = 160 = (v)/(0.35) \rightarrow v = 56 \text{ m/yr}$$

d. Use Eq. III.1.1 to find the hydraulic conductivity:

$$v = ki = (k)(0.02) = 56 \text{ m/yr} \rightarrow k = 2800 \text{ m/yr} = 7.7 \text{ m/d}$$

Discussion. The neighborhood pumping during the drought would increase the natural groundwater gradient. During the pumping period, the groundwater moved faster, and so did the plume. This resulted in a larger size of the plume. In other words, the plume would have traveled a shorter distance without the pumping. The hydraulic conductivity of the aquifer would be smaller than this estimate, 7.7 m/d.

Example III.4.4C Retardation factor and partition of contaminants

The toluene concentration of the groundwater in a contaminated aquifer was determined to be 500 ppb. Assuming no free product phase present, estimate the partition of toluene in the two phases, i.e., dissolved in liquid and adsorbed onto the solid phases.

From the RI work, the following parameters were determined:

Retardation factor = 4.0
Porosity = 0.35
Bulk density of the aquifer matrix = 1.8 g/cm^3

Strategy. To determine the partition between the liquid phase and solid phase, we need to know the partition coefficient. The partition coefficient can be found from the retardation factor.

Solution:

a. Use Eq. III.4.12 to determine the partition coefficient, K_p:

$$R = 1 + \frac{\rho_b K_p}{\phi} = 4 = 1 + \frac{(1.8)K_p}{0.35}$$

So,

$$K_p = 0.583 \text{ L/kg}$$

Use Eq. II.3.11 to find the contaminant concentration in the soil, X

$$X = K_p C = (0.583)(0.5) = 0.292 \text{ mg/kg}$$

b. Basis: 1 L of soil

$$\text{Mass dissolved in liquid} = (V)(\phi)(C) = (1 \text{ L})(0.35)(0.5 \text{ mg/kg}) = 0.175 \text{ mg}$$

$$\text{Mass adsorbed on the solid} = (V)(\rho_b)(X) = (1 \text{ L})(1.8 \text{ kg/L})(0.292 \text{ mg/kg}) = 0.526 \text{ mg}$$

$$\% \text{ mass in liquid} = (0.175) \div [(0.175) + (0.526)] = 25\%$$

Discussion. This example illustrates that the majority of the contaminant was attached to the solids; only 25% was in the dissolved phase. This partially explains why the cleanup takes a long time for groundwater remediation using the pump-and-treat method.

III.5 Contaminant transport in the vadose zone

Contaminant travel in the vadose zone can occur in three ways: (1) volatilizing into the air void and traveling as vapor, (2) becoming dissolved into the soil moisture and/or into the infiltrating water and then traveling with the liquid, and (3) moving downward by gravity as the immiscible phase. This section describes these transport pathways.

III.5.1 Liquid movement in the vadose zone

Liquid flow through the vadose zone can be described by a differential equation, and its one-dimensional form is

$$\frac{\partial}{\partial z}\left[K\frac{\partial \Psi}{\partial z}\right]+\frac{\partial K}{\partial z}=\frac{\partial \theta_w}{\partial \Psi}\frac{\partial \Psi}{\partial t} \qquad \text{[Eq. III.5.1]}$$

where K is the hydraulic conductivity, θ_w is the volumetric water content, ψ is the soil water pressure head (the sum of the gravity potential and the moisture potential), and t is time. The major differences between this equation and the equation for one-dimensional groundwater flow (i.e., Darcy's Law) are (1) the hydraulic conductivity in the vadose zone is a function of ψ, and hence of θ_w, and (2) the pressure head is a function of time. These make Eq. III.5.1 nonlinear, time dependent, and more difficult to solve than the simple Darcy's equation. (If K is a constant and pressure is independent of time, then Eq. III.5.1 can be simplified to the Darcy's equation.)

The hydraulic conductivity of a vadose zone is largest at water saturation and decreases as the water content decreases. As the moisture content decreases, air occupies most of the pore void and leaves less cross-sectional area for water transport. Consequently, the hydraulic conductivity decreases. At very low moisture content, the water film covering the soil particles becomes very thin. The attractive forces between the water molecules and the soil particles become so strong that no water will move. At this point, the hydraulic conductivity is approaching zero. The hydraulic conductivity at a given moisture can be found from the relative permeability for that moisture, k_r (a dimensionless term), and the hydraulic conductivity at saturation, K_s, as

$$K = k_r K_S \qquad \text{[Eq. III.5.2]}$$

The relative hydraulic conductivity varies from 1.0 at 100% saturation to 0.0 at 0% saturation.

The contaminant transport in the dissolved phase in the vadose zone can be described by an advection–dispersion equation, and its one dimensional form is

$$\frac{\partial(\theta_w C)}{\partial t} = \frac{\partial^2(\theta_w DC)}{\partial z^2} - \frac{\partial(\theta_w vC)}{\partial z} \pm RXNs \qquad \text{[Eq. III.5.3]}$$

This equation is similar to the one for the saturated zone, except the soil moisture content, θ_w, is a variable and the velocity and dispersion coefficient depend on the moisture content. The dispersion coefficient is analogous to the dispersion term in the saturated zone, except v is a function of the moisture content, as

$$D = D_d + D_h = \xi D_o + \alpha v(\theta_w) \qquad \text{[Eq. III.5.4]}$$

Example III.5.1 Estimate the hydraulic conductivity in the vadose zone

A subsurface soil is relatively sandy and has a hydraulic conductivity of 500 gpd/ft^2 when the soil is saturated. Estimate its hydraulic conductivity (a) when the water saturation is 40% and (b) when the water saturation is 90%. The relative permeability for sand at 40% saturation is 0.02, and that at 90% saturation is 0.44.

Solution:

a. Use Eq. III.5.2 to find the hydraulic conductivity at 40% saturation:

$$K = (0.02)(500) = 10 \text{ gpd/ft}^2$$

b. Use Eq. III.5.2 to find the hydraulic conductivity at 90% saturation:

$$K = (0.44)(500) = 220 \text{ gpd/ft}^2$$

Discussion. At 40% water saturation, the hydraulic conductivity is close to zero, and, at 90% water saturation, the hydraulic conductivity is 44% of the maximum value. The water saturation is the percentage of the pore space that is occupied by the water: 100% for saturated soil and 0% for dry soil.

III.5.2 Gaseous diffusion in the vadose zone

Under nonpumping conditions, the molecular diffusion is the prime mechanism for gas-phase transport. The transport equation can be expressed by Fick's Law, and its one dimensional form is

$$\xi_a \phi_a D \frac{\partial^2 G}{\partial x^2} = \frac{\partial(\phi_a G)}{\partial t} \qquad \text{[Eq. III.5.5]}$$

where D is the free-air diffusion coefficient, G is the contaminant concentration in the gas phase, ϕ_a is the air-filled porosity, and ξ_a is the air-phase tortuosity factor. The ξ_a term accounts for the diffusion taking place within a porous medium rather than in an open air space. It can be estimated from empirical equations such as the Millington–Quirk equation.[8]

$$\xi_a = \frac{\phi_a^{2.333}}{\phi_t^2} \qquad \text{[Eq. III.5.6]}$$

where ϕ_t is the total porosity, which is the sum of the air-filled porosity and the volumetric water content ($\phi_t = \phi_a + \phi_w$). The air-phase tortuosity factor varies from zero, when the entire pore space is occupied by water (saturated condition), to about 0.8, when the porosity is high and the medium is dry.

The values of the free air diffusion coefficient for selected compounds can be found in Table II.3.C. The free air diffusion coefficient is generally 10,000 times higher than that in a dilute aqueous solution. The diffusion coefficient can also be estimated from the diffusion coefficient of another compound of similar species and molecular weight by the following relationship (same as that for liquid as in Eq. III.4.6):

$$\frac{D_1}{D_2} = \sqrt{\frac{\text{MW}_2}{\text{MW}_1}} \qquad \text{[Eq. III.5.7]}$$

The diffusion coefficient is inversely proportional to the square root of its molecular weight. The heavier the contaminant, the more difficult it is to diffuse through the air. Temperature can have an influence on the diffusion coefficient. The diffusion coefficient increases with temperature, and the following relationship applies:

$$\frac{D_o @ T_1}{D_o @ T_2} = \left(\frac{T_1}{T_2}\right)^m \qquad \text{[Eq. III.5.8]}$$

where T is the temperature in Kelvin. Theoretically, the exponent, m, should be 1.5; however, experimental data indicate that it ranges from 1.75 to 2.0.

Example III.5.2A Estimate the air-phase tortuosity factor

A subsurface soil is relatively sandy and has a porosity of 0.45. Estimate its air-phase tortuosity factor:

a. When the volumetric water content is 0.3
b. When the volumetric water content is 0.05

Solution:
a. For $\phi_w = 0.3$ and $\phi_t = 0.45$,

$$\phi_a = 0.45 - 0.3 = 0.15$$

Use Eq. III.5.2 to find the air-phase tortuosity factor at $\phi_w = 0.3$:

$$\xi_a = \frac{(0.15)^{2.333}}{(0.45)^2} = 0.059$$

b. For $\phi_w = 0.05$ and $\phi_t = 0.45$,

$$\phi_a = 0.45 - 0.05 = 0.40$$

Use Eq. III.5.2 to find the air-phase tortuosity factor at $\phi_w = 0.05$

$$\xi_a = \frac{(0.40)^{2.333}}{(0.45)^2} = 0.58$$

Discussion. For this case, the air-phase tortuosity is approximately ten times higher when the volumetric water content drops from 0.3 to 0.05. The volumetric water content here is the percentage of total soil volume (not the void volume) occupied by water.

Example III.5.2B *Estimate the diffusion coefficient at different temperatures*

The diffusion coefficient of benzene in dilute aqueous solution at 20°C is 1.02×10^{-5} cm²/s (Table III.4.A), and the free-air diffusion coefficient of benzene is 0.092 cm²/s at 25°C (Table II.3.C). Use these reported values to estimate:

a. The ratio of diffusion coefficient of benzene in free air and in dilute aqueous solution at 20°C
b. The free air diffusion coefficient of toluene at 20°C

Solution:
a. Use Eq. III.5.8 and $m = 2$ (assumed) to determine the free-air diffusion coefficient of benzene at 20°C:

$$\frac{0.092}{D_o @ T_2} = \left(\frac{298}{293}\right)^2$$

So, the free-air diffusion coefficient of benzene at 20°C = 0.089 cm²/s. The ratio between the free-air and liquid diffusion coefficients

$$(0.089) \div 1.02 \times 10^{-5} = 8720$$

b. The molecular weight of toluene ($C_6H_5CH_3$) is 92, and the molecular weight of benzene (C_6H_6) is 78. Use Eq. III.5.7 to determine the diffusion coefficient:

$$\frac{D_1}{D_2} = \frac{(0.083)}{D_2} = \sqrt{\frac{92}{78}}$$

So, the diffusion coefficient of toluene at 20°C = 0.082 cm²/s.

Discussion

1. The diffusion coefficient of benzene in free air is 8720 higher than in the dilute aqueous phase.
2. The diffusion coefficient of toluene estimated from that of benzene and the molecular weight relationship, 0.082 cm²/s, is essentially the same as that in Table II.3.C, 0.083 cm²/s.

III.5.3 Retardation factor for vapor migration in the vadose zone

For a contaminated air stream flowing through a porous medium, the gas-phase retardation factor can be derived as[8]

$$R_a = 1 + \frac{\rho_b K_p}{\phi_a H} + \frac{\phi_w}{\phi_a H} \quad \text{[Eq. III.5.9]}$$

This retardation factor will be a constant if the water content, ϕ_w, does not change. It is analogous to the retardation factor, R, for the movement of contaminant in the aquifer. The movement of the contaminant in the vadose zone will be retarded by a factor of R_a. The second term represents the partitioning from the vapor phase, through the soil moisture, to the solid phase. The third term on the right-hand side of Eq. III.5.9 represents the partitioning of the contaminant between the vapor phase and the soil moisture phase. As the contaminants in the vapor phase move through the air-filled pores, the migration rate of the contaminant in the air is less than that of the air itself because of the loss of mass to the soil moisture and to the soil organic carbon. Under the condition of no advective flow, the gas-phase retardation factor can be defined as the ratio of the diffusion rate of an inert compound such as nitrogen to the diffusion rate of the contaminant. Under advective flow, it can be used as the relative measure to compare the migration rates of compounds with different retardation factors. For soil venting applications, the air-phase retardation factor is also the minimum number of pore volumes that must pass through the contaminated zone to clean up the zone. It is considered as the minimum because this approach ignores the effects of mass transfer limitations among the phase, subsurface heterogeneity, and unequal travel time from the outer edge of the plume to the vapor extraction well.[8]

As shown in Eq. III.5.9, the air-phase retardation factor increases with soil moisture content and K_p but decreases with Henry's constant. A higher moisture content means a larger water reservoir to retain the contaminants, and a larger K_p value indicates that the soil has a larger organic content or the contaminant is more hydrophobic. On the other hand, compounds with high Henry's constant values have a stronger tendency to volatilize into the pores. The Henry's constant increases with increasing temperature and, thus, a smaller air-phase retardation factor. Therefore, for a soil venting application, fewer pore volumes of air need to be moved through the contaminated zone to remove the contaminant at higher temperature.

Example III.5.3 *Determination of the air-phase retardation factor*

The vadose zone underneath a landfill is contaminated by landfill leachates containing benzene, 1,2-dichloroethane (DCA), and pyrene.

Estimate the air-phase retardation factor using the following data from the site assessment:

Vadose zone soil porosity = 0.40
Volumetric water content = 0.15
Bulk density of soil = 1.8 g/cm^3
Fraction of organic carbon of soil = 0.015
Temperature of the formation = 25°C
$K_{oc} = 0.63\ K_{ow}$

Solution:

a. From Table II.3.C,

$$H = 5.55 \text{ atm/M for benzene (at 25°C)}$$

Use Table II.3.B to covert it to a dimensionless value:

$$H^* = H/RT = (5.55)/[(0.082)(298)] = 0.227$$

Similarly, for 1,2-DCA (Henry's constant value in the table is for 20°C; we use this value for 25°C as an approximate value) and pyrene,

$$H^* = H/RT = (0.98)/[(0.082)(298)] = 0.04 \text{ (for 1,2 DCA)}$$

$$H^* = H/RT = (0.005)/[(0.082)(298)] = 0.0002 \text{ (for pyrene)}$$

b. From Example III.4.3,

$$K_p = (0.015)(85) = 1.275 \text{ (for benzene)}$$

$$K_p = (0.015)(22) = 0.32 \text{ (for 1,2-DCA)}$$

$$K_p = (0.015)(47{,}800) = 717 \text{ (for pyrene)}$$

c. Use Eq. III.5.9 to find the air-phase retardation factor:

$$R_a = 1 + \frac{\rho_b K_p}{\phi_a H} + \frac{\phi_w}{\phi_a H}$$

$$= 1 + \frac{(1.8)(1.275)}{(0.25)(0.227)} + \frac{(0.15)}{(0.25)(0.227)} = 44 \text{ for benzene}$$

$$R_a = 1 + \frac{\rho_b K_p}{\phi_a H} + \frac{\phi_w}{\phi_a H}$$

$$= 1 + \frac{(1.8)(0.32)}{(0.25)(0.04)} + \frac{(0.15)}{(0.25)(0.04)} = 73.6 \text{ for 1,2-DCA}$$

$$R_a = 1 + \frac{\rho_b K_p}{\phi_a H} + \frac{\phi_w}{\phi_a H}$$

$$= 1 + \frac{(1.8)(717)}{(0.25)(0.0002)} + \frac{(0.15)}{(0.25)(0.0002)} = 2.6 \times 10^7 \text{ for pyrene}$$

Discussion. Pyrene is very hydrophobic and has a low Henry's constant. Its air-phase retardation factor is much higher than that of benzene or 1,2-DCA.

References

1. Driscoll, F. G., *Groundwater and Wells,* 2nd ed., Johnson Division, St. Paul, MN, 1986.
2. Fetter, C.W., Jr., *Applied Hydrogeology,* Charles E. Merrill Publishing, Columbus, OH, 1980.
3. Freeze, R. A. and Cherry, R. A., *Groundwater,* Prentice Hall, Englewood Cliffs, NJ, 1979.

4. Sherwood, T. K., Pigford, R. L., and Wilke, C. R., *Mass Transfer,* McGraw-Hill, New York, 1975.
5. Todd, D. K., *Groundwater Hydrology,* 2nd ed., John Wiley & Sons, New York, 1980.
6. U.S. EPA, Transport and Face of Contaminants in the Subsurface, EPA/625/4-89/019, U.S. EPA, Washington, DC, 1989.
7. U.S. EPA, Ground Water Volume I: Ground Water and Contamination, EPA/625/6-90/016a, U.S. EPA, Washington, DC, 1990.
8. U.S. EPA, Site Characterization for Subsurface Remediation, EPA/625/4-91/026, U.S. EPA, Washington, DC, 1991.

chapter four

Mass balance concept and reactor design

Various treatment processes are employed in remediation of contaminated soil or groundwater. Treatment processes are generally classified as physical, chemical, biological, and thermal processes. Treatment systems often consist of a series of unit operations/processes, which form a process train. Each unit operation/process contains one or more reactors. A reactor can be considered as a vessel in which the processes occur. Environmental engineers are often in charge of or, at least, participate in preliminary design of the treatment system. Basically, the preliminary design involves selection of treatment processes and reactor type as well as sizing the reactors.

For treatment system design, treatment processes should be selected first by screening the alternatives. Many factors should be considered in selection of treatment processes. Common selection criteria are implementability, effectiveness, cost, and regulatory consideration. In other words, an optimum process would be the one that is implementable, effective in removal of contaminants, cost efficient, and in compliance with the regulatory requirements.

Once the treatment processes are selected for a remediation project, engineers will then design the reactors. Preliminary reactor design usually includes selecting appropriate reactor types, sizing reactors, and determining the number of reactors needed and their optimal configuration. To size the reactors, engineers first need to know if the desirable reactions or activities would occur in the reactors and what the optimal operating conditions such as temperature and pressure would be. Information from chemical thermodynamics, or more practically a pilot study, would provide the answers to these questions. If the desired reactions are feasible, the engineers then need to determine the rates of these reactions, which is a subject of chemical kinetics. The reactor size is then determined, based on mass loading to the reactor, reaction rate, and type of reactor.

This chapter introduces the mass balance concept, which is the basis for process design. Then it presents reaction kinetics as well as types, configuration, and sizing of reactors. From this chapter you will learn how to determine the rate constant, removal efficiency, optimal arrangement of reactors, required residence time, and reactor size for your specific applications.

IV.1 Mass balance concept

The mass balance (or material balance) concept serves as a basis for designing environmental engineering systems (reactors). The mass balance concept is nothing but conservation of mass. Matter can neither be created nor destroyed, but it can be changed in form (a nuclear process is one of the few exceptions). The fundamental approach is to show the changes occurring in the reactor by the mass balance analysis. The following is a general form of a mass balance equation:

$$\begin{bmatrix}\text{Rate of mass}\\ \text{ACCUMULATED}\end{bmatrix} = \begin{bmatrix}\text{Rate of mass}\\ \text{IN}\end{bmatrix} - \begin{bmatrix}\text{Rate of mass}\\ \text{OUT}\end{bmatrix} \pm \begin{bmatrix}\text{Rate of mass}\\ \text{GENERATED or}\\ \text{DESTROYED}\end{bmatrix}$$

[Eq. IV.1.1]

Performing a mass balance on an environmental engineering system is just like balancing your checkbook. The rate of mass accumulated (or depleted) in a reactor can be viewed as the rate that money is accumulated (or depleted) in your checking account. How fast the balance changes depends on how much and how often the money is deposited and/or withdrawn (rate of mass input and output), interest accrued (rate of mass generated), and bank charges for monthly service and ATM fees imposed (rate of mass destroyed).

In using the mass balance concept to analyze an environmental engineering system, one usually begins by drawing a process flow diagram and employing the following procedure:

Step 1: Draw system boundaries or boxes around the unit processes/operations or flow junctions to facilitate calculations.

Step 2: Place known flow rates and concentrations of all streams, sizes and types of reactors, and operating conditions such as temperature and pressure on the diagram.

Step 3: Calculate and convert all known mass inputs, outputs, and accumulation to the same units and place them on the diagram.

Step 4: Mark unknown (or the ones to be solved) inputs, outputs, and accumulation on the diagram.

Step 5: Perform the necessary analyses/calculations using the procedures described in this chapter.

A few special cases or reasonable assumptions would simplify the general mass balance equation, Eq. IV.1.1, and make the analysis easier. Three common ones are presented below:

a. *No Reactions Occurring:* If the system has no chemical reactions occurring, such as a mixing process, there is no increase or decrease of compound mass due to reactions. The mass balance equation would become

$$\begin{bmatrix}\text{Rate of mass}\\ \text{ACCUMULATED}\end{bmatrix} = \begin{bmatrix}\text{Rate of mass}\\ \text{IN}\end{bmatrix} - \begin{bmatrix}\text{Rate of mass}\\ \text{OUT}\end{bmatrix} \qquad \text{[Eq. IV.1.2]}$$

b. *Batch Reactor:* For a batch reactor, there is no input into or output out of the reactor. The mass balance equation can be simplified into

$$\begin{bmatrix}\text{Rate of mass}\\ \text{ACCUMULATED}\end{bmatrix} = \pm \begin{bmatrix}\text{Rate of mass}\\ \text{GENERATED or}\\ \text{DESTROYED}\end{bmatrix} \qquad \text{[Eq. IV.1.3]}$$

Examples of using Eq. IV.1.3 will be provided in later sections of this chapter.

c. *Steady-State Conditions:* To maintain the stability of treatment processes, treatment systems are usually kept under steady-state conditions after a start-up period. A steady-state condition basically means that flow and concentrations at any location within the treatment process train are not changing with time. Although the concentration and/or flow rate of the influent waste stream entering a soil/groundwater system typically fluctuate, engineers may want to incorporate devices such as equalization tanks to dampen the fluctuation. This is especially true for treatment processes that are very sensitive to fluctuation of mass loading (biological processes are good examples).

For a reactor under a steady-state condition, although reactions are occurring inside the reactor, the rate of mass accumulation in the reactor would be zero. Consequently, the left-hand side term of Eq. IV.1.1 becomes zero. The mass balance equation can then be reduced to

$$0 = \begin{bmatrix}\text{Rate of mass}\\ \text{IN}\end{bmatrix} - \begin{bmatrix}\text{Rate of mass}\\ \text{OUT}\end{bmatrix} \pm \begin{bmatrix}\text{Rate of mass}\\ \text{GENERATED or}\\ \text{DESTROYED}\end{bmatrix} \qquad \text{[Eq. IV.1.4]}$$

Assumption of steady-state is frequently used in the analysis of flow reactors, and examples of using Eq. IV.1.4 will be provided in later sections of this chapter.

The general mass balance equation, Eq. VI.1.1, can also be expressed as

$$V\frac{dC}{dt} = \sum Q_{in}C_{in} - \sum Q_{out}C_{out} \pm (V \times \gamma) \qquad \text{[Eq. IV.1.5]}$$

where V is the volume of the system (reactor), C is the concentration, Q is the flow rate, and γ is the reaction rate. The following sections will demonstrate the role of the reaction in the mass balance equation and how it affects the reactor design.

Example IV.1.1 *Mass balance equation — air dilution (no chemical reaction occurring)*

A glass bottle containing 900 mL of methylene chloride (CH_2Cl_2, specific gravity = 1.335) was accidentally left uncapped in a poorly ventilated room (5 m × 6 m × 3.6 m) over a weekend. On the following Monday it was found that two thirds of methylene chloride had volatilized. For a worst-case scenario, would the concentration in the room air exceed the permissible exposure limit (PEL) of 100 ppmV?

An exhaust fan (Q = 200 ft^3/min) was turned on to vent the fouled air in the laboratory. How long will it take to reduce the concentration down below the PEL?

Stategy. This is a special case (no reactions occurring) of the general mass balance equation. For this case Eq. IV.1.5 can be simplified into

$$V\frac{dC}{dt} = \sum Q_{in}C_{in} - \sum Q_{out}C_{out} \qquad \text{[Eq. IV.1.6]}$$

The equation can be further simplified with the following assumptions:

1. The air leaving the laboratory is only through the exhaust fan and the air ventilation is equal to the rate of air entering the laboratory ($Q_{in} = Q_{out} = Q$)
2. The air entering the laboratory does not contain methylene chloride ($C_{in} = 0$).
3. The air in the laboratory is fully mixed, thus the concentration of methylene chloride in the laboratory is uniform and is the same as that of the air vented by the fan ($C = C_{out}$).

$$V\frac{dC}{dt} = -QC \qquad \text{[Eq. IV.1.7]}$$

It is a first-order differential equation. It can be integrated with initial condition, $C = C_0$ at $t = 0$:

$$\frac{C}{C_o} = e^{-(Q/V)t} \quad \text{or} \quad C = C_o e^{-(Q/V)t} \qquad \text{[Eq. IV.1.8]}$$

Solution:

a. Methylene chloride concentration in the laboratory before ventilation can be found as 2101 ppmV (see Example II.1.1C for detailed calculations).
b. The size of the reactor, V = the size of the laboratory

$$= (5\text{ m})(6\text{ m})(3.6\text{ m}) = 108\text{ m}^3.$$

The system flow rate, Q = ventilation rate

$$= 200\text{ ft}^3/\text{min} = (200\text{ ft}^3/\text{min}) \div (35.3\text{ ft}^3/\text{m}^3) = 5.66\text{ m}^3/\text{min}.$$

The initial concentration, C_0 = 2101 ppmV. The final concentration, C = 100 ppmV.

$$100 = (2101)e^{-(5.66/108)t}$$

Thus, t = 58 min.

Discussion. The actual time required would be longer than 58 minutes because the assumption of complete mix inside the room may not be valid.

IV.2 Chemical kinetics

Chemical kinetics is concerned with the rate at which chemical reactions occur. This section discusses the rate equation, reaction rate constant, and reaction order. Half-life, a term commonly used with regard to the fate of contaminants in the environment, is also described.

IV.2.1 Rate equations

In addition to the mass balance concept, the other relationship required for design of a homogeneous reactor is the reaction rate equation. The following general mathematical expression describes the rate that the concentration of species A, C_A, is changing with time:

$$\gamma_A = \frac{dC_A}{dt} = -kC_A^n \qquad \text{[Eq. IV.2.1]}$$

where n is the reaction order, k is the reaction rate constant, and γ_A is the rate of conversion of species A. If the reaction order, n, is equal to 1, it is called a first-order reaction. It implies that the reaction rate is proportional to the concentration of the species. In other words, the higher the compound concentration, the faster the reaction rate. The first-order kinetics is applicable for many environmental engineering applications. Consequently, discussions in this book are focused on the first-order reactions and their applications. The first-order reaction can then be written as

$$\gamma_A = \frac{dC_A}{dt} = -kC_A \qquad \text{[Eq. IV.2.2]}$$

The rate constant itself provides valuable information regarding the reaction. A larger k value implies a faster reaction rate, which, in turn, demands a smaller reactor volume in order to achieve a specified conversion. The value of k varies with temperature. In general, the higher the temperature, the larger the k value will be for a reaction.

What would be the units of the reaction rate constant for a first-order reaction? Let us take a close look at Eq. IV.2.2. In that equation the unit for dC_A/dt is concentration/time and that of C is concentration; therefore, the unit of k should be 1/time. Consequently, if a reaction rate is given as 0.25 d^{-1}, the reaction should be a first-order reaction. The units of k for zeroth-order reactions and second-order reactions should be [(concentration)/time] and $[(\text{concentration})(\text{time})]^{-1}$, respectively.

Eq. IV.2.2 tells us that the concentration of compound A is changing with time. This equation can be integrated between $t = 0$ and time t:

$$\ln\frac{C_A}{C_{A0}} = -kt \quad or \quad \frac{C_A}{C_{A0}} = e^{-kt} \qquad \text{[Eq. IV.2.3]}$$

where C_{A0} is the concentration of compound A at $t = 0$, and C_A is the concentration at time t.

Example IV.2.1A Estimate the rate constant from two known concentration values

An accidental gasoline spill occurred at a site 5 days ago. The TPH concentration at a specific location in soil dropped from an initial 3000 mg/kg to the current 2550 mg/kg. The decrease in concentration is mainly attributed to natural biodegradation and volatilization. Assume that both removal mechanisms are first-order reactions and the reaction rate constants for both mechanisms are independent of contaminant concentration and are constant. Estimate how long it will take for the concentration to drop below 100 mg/kg by these natural attenuation processes.

Strategy. Only the initial concentration and the concentration at day 5 are given. We need to take a two-step approach to solve the problem: first determine the rate constant and then use the rate constant to determine the time needed to reach a final concentration of 100 mg/kg.

Two removal mechanisms are occurring at the same time, they are both first order. They can be represented by one single equation and one combined rate constant.

$$\frac{dC}{dt} = -k_1C - k_2C = -(k_1 + k_2)C = -kC \qquad \text{[Eq. IV.2.4]}$$

Solution:

a. Insert the initial concentration and the concentration at day 5 into Eq. IV.2.3 to obtain *k:*

$$\ln\frac{2550}{3000} = -k(5)$$

So,

$$k = 0.0325/\text{d}$$

b. For the concentration to drop below 100 mg/kg, it will take (from Eq. IV.2.3):

$$\ln\frac{100}{3000} = -0.0325(t)$$

$$t = 105 \text{ days}$$

Example IV.2.1B *Estimate the rate constant from two known concentration values*

The soil of a subject site was contaminated by an accidental spill of gasoline. A soil sample, taken 10 days after removal of the polluting source, showed a concentration of 1200 mg/kg. The second sample taken at 20 days showed a drop of concentration, at 800 mg/kg. Assuming that a combination of all the removal mechanisms including volatilization, biodegradation, and oxidation show first-order kinetics, estimate how long it will take for the concentration to drop below 100 mg/kg without any remediation measures taken.

Strategy. Two concentrations at two different time steps are given. We should take a two-step approach to solve the problem. We need to determine the initial concentration and *k* first (two equations for the two unknowns).

Solution:

a. Determine the initial concentration (immediately after the spill) and k. At t = 10 days, insert the concentration value into Eq. IV.2.3

$$\frac{1200}{C_i} = e^{-k(10)}$$

At t = 20 days, insert the concentration value into Eq. IV.2.3

$$\frac{800}{C_i} = e^{-k(20)}$$

Dividing both sides of the first equation by the corresponding sides of the second equation, we can obtain

$$\frac{1200}{800} = 1.5 = e^{-10k} \div e^{-25k} = e^{-10k-(-25k)} = e^{15k}$$

Thus, k = 0.027/d.

Then C_i can be easily determined by inserting the value of k into either of the first two equations:

$$\frac{1200}{C_i} = e^{-(0.027)(10)} = 0.763$$

So,

$$C_i = 1572 \text{ mg/kg}$$

b. For the concentration to drop below 100 mg/kg, it will take

$$\frac{100}{1572} = 0.0636 = e^{-0.027t}$$

$$t = 102 \text{ days}$$

IV.2.2 *Half-life*

The half-life can be defined as the time it takes to convert one-half of the compound of concern. For the first-order reaction, the half-life (often shown as $t_{1/2}$) can be found from Eq. IV.2.3 by substituting $C_{A,t}$ by one half of $C_{A,0}$, i.e., $(0.5)(C_{A,0})$,

$$t_{1/2} = \frac{\ln 2}{k} = \frac{0.693}{k} \quad \text{[Eq. IV.2.5]}$$

Example IV.2.2A Half-life calculation

The half-life of 1,1,1-trichloroethane (1,1,1-TCA) in a subsurface environment was determined to be 180 days. Assume that all the removal mechanisms are first order. Determine (1) the rate constant and (2) the time needed to drop the concentration down to 10% of the initial concentration.

Solution:

a. The rate constant can be easily determined from Eq. IV.2.5 as

$$t_{1/2} = 180 = \frac{0.693}{k}$$

Thus,

$$k = (3.85 \times 10^{-3})/\mathrm{d}$$

b. Use Eq. IV.2.3 to determine the time needed to drop the concentration down to 10% of the initial (i.e., $C = 0.1 \times C_i$).

$$\frac{C}{C_i} = \frac{1}{10} = e^{-(3.85E-3)(t)}$$

Therefore,

$$t = 598 \text{ days}$$

Example IV.2.2B Half-life calculation

In some occasions, the decay rate is expressed as T_{90} instead of $t_{1/2}$. T_{90} is the time required for 90% of the compound to be converted (or the concentration to drop to 10% of the initial value). Derive an equation to relate T_{90} with the first-order reaction rate constant.

Solution:

The relationship between T_{90} and k can be determined from Eq. IV.2.3 as

$$\frac{C}{C_i} = \frac{1}{10} = e^{-kT_{90}}$$

Then,

$$T_{90} = \frac{-\ln(0.1)}{k} = \frac{2.30}{k} \qquad \text{[Eq. IV.2.6]}$$

IV.3 Types of reactors

Reactors are typically classified based on their flow characteristics and the mixing conditions within the reactor. Reactors may be operated in either a batchwise or a continuous-flow mode. In a batch reactor, the reactor is charged with the reactants, the contents are well mixed and left to react, and then the resulting mixture is discharged. The batch reactor is considered as an unsteady-state reactor because the composition of the reactor content changes with time. The capital cost of a batch reactor is usually less that that of a continuous-flow reactor, but it is very labor intensive and operating costs are much higher. It is usually limited to small installations and to cases when raw materials are expensive.

In a continuous-flow reactor, the feed to the reactor and the discharge from it are continuous. In most of the cases the flow reactors are operated under steady-state conditions in which the feed stream flow rate, its composition rate, the reaction condition in the reactor, and the withdrawal rate are constant with respect to time. Frequently, reaction kinetics are studied in the laboratory using a batch reactor to determine the rate constant, k. The application of the kinetic constant, k, to the design of a continuous-flow reactor, however, involves no changes in kinetic principles; thus, it is valid. In general, there are two ideal types of flow reactors: continuous flow stirred tank reactor (CFSTR) and plug flow reactor (PFR). They are classified by the mixing conditions within the reactors.

The CFSTR consists of a stirred tank that has a feed stream(s) of the reactants and a discharge stream(s) of reacted materials. The CFSTR is usually round, square, or slightly rectangular in plan view, and it is necessary to provide sufficient mixing. The stirring of a CFSTR is extremely important, and it is assumed that the fluid in the reactor is perfectly mixed, that is, the contents are uniform throughout the reactor volume. As a result of mixing, the composition of the discharge stream(s) is the same as that of the reactor contents. Therefore, it is also called a completely stirred tank reactor (CSTR) or completely mixed flow reactor (CMF). Under steady-state conditions, the effluent concentration and concentration at any location within the reactor are the same and should not change with time.

The PFR ideally has the geometric shape of a long tube or tank and has a continuous flow in which the fluid particles pass through the reactor in series. The reactants enter at the upstream end of the reactor, and the products leave at the downstream end. Ideally, there is no induced mixing between elements of fluid along the direction of flow. Those fluid particles that enter the reactor first will leave first. The composition of the reacting fluid changes in the direction of flow. For the case of contaminant removal or destruction, the concentration will be the highest at the entrance and dropped continuously to the effluent value at the exit condition. Under steady-state conditions, the effluent concentration and concentration at any location within the reactor should not change with time.

It should be noted that CFSTRs and PFRs are ideal reactors. The continuous flow reactors in the real world behave somewhere between these ideal cases.

IV.3.1 Batch reactors

Let us consider a batch reactor with a first-order reaction. By combining Eq. IV.2.2 and Eq. IV.1.3, the mass balance equation can be expressed as:

$$V\frac{dC}{dt} = (V \times \gamma) = V(-kC)$$

$$\text{or } \frac{dC}{dt} = -kC \qquad \text{[Eq. IV.3.1]}$$

It is a first-order differential equation, and it can be integrated with the initial condition ($C = C_i$ at $t = 0$) and the final condition (C = final concentration, C_f at t = residence time, τ). The residence time, τ, can be defined as the time that the fluid stays inside the reactor and undergoes reaction. The integral of Eq. IV.3.1 is

$$\frac{C_f}{C_i} = e^{-k\tau} \quad \text{or} \quad C_f = (C_i)e^{-k\tau} \qquad \text{[Eq. IV.3.2]}$$

Table IV.3.A summarizes the design equations for batch reactors in which zeroth-, first-, and second-order reactions take place.

Table IV.3.A Design Equations for Batch Reactors

Reaction order, n	Design equation	
0	$C_f = C_i - k\tau$	[Eq. IV.3.3]
1	$C_f = C_i(e^{-k\tau})$	same as [Eq. IV.3.2]
2	$C_f = \dfrac{C_i}{1+(k\tau)C_i}$	[Eq. IV.3.4]

Example IV.3.1A *Batch reactor (determine the required residence time with known rate constant)*

A batch reactor is to be designed to treat soil contaminated with 200 mg/kg of PCBs. If the required removal, conversion, or reduction of PCBs is 90% and the rate constant is 0.5 hr^{-1}, what is the required residence time for the batch reactor? What is the required residence time if the desired final concentration is 10 mg/kg?

Strategy

1. There are four parameters in the equation for the batch reactor: two concentrations, k, and time. We need to know three of the four to determine the one left.
2. Although the order of the reaction is not mentioned in the problem statement, it is assumed to be a first-order reaction because the dimensions of k are [1/(time)].

Solution:

a. For a 90% reduction (η = 90%):

$$C_f = C_i\ (1 - \eta) = 200\ (1 - 90\%) = 20 \text{ mg/kg}$$

Insert the known values into Eq. IV.3.2,

$$\frac{20}{200} = 0.1 = e^{-(0.5)t}$$

$$\tau = 4.6 \text{ hr}$$

b. To achieve a final concentration of 10 mg/kg:

$$\frac{10}{200} = 0.05 = e^{-(0.5)t}$$

$$\tau = 6.0 \text{ hr}$$

Example IV.3.1B *Batch reactor (determine the required residence time with unknown rate constant)*

A batch reactor was installed to remediate PCB-contaminated soil. A test run was conducted with an initial PCB concentration of 250 mg/kg. After 10 hours of batchwise operation, the concentration was dropped to 50 mg/kg. However, it is required to reduce the concentration down to 10 mg/kg. Determine the required residence time to achieve the final concentration of 10 mg/kg.

Strategy. It requires a two-step approach to solve the problem. The first is to determine the rate constant using the given information. Then, use this obtained k value to estimate the residence time for other conversions. The given information did not tell us the order of the reaction. We assume that it is a first-order reaction. This should be confirmed with additional test data.

Solution:

a. Insert the known values into Eq. IV.3.2 to find the value of k:

$$\frac{50}{250} = 0.20 = e^{-k(10)}$$

$$k = 0.161\ \text{hr}^{-1}$$

b. The time required to achieve a concentration of 10 mg/kg:

$$\frac{10}{250} = 0.04 = e^{-0.161t}$$

$$\tau = 20.0\ \text{hours}$$

Discussion. It is assumed that the first-order reaction applies in this calculation. One should check the validity of this assumption, for example, by running the pilot-run longer or running a bench-scale batchwise experiment. For example, if the run is extended to 20 hours and the final concentration is close to 10 mg/kg, the assumption of first-order kinetics should be valid.

Example IV.3.1C Determine the rate constant from batch experiments

An in-vessel bioreactor is designed to remediate soils contaminated with cresol. A bench-scale batch reactor was set up to determine the reaction order and rate constant. The following concentrations of cresol in the batch reactor at various times were observed and recorded as

Time (hours)	Cresol concentration (mg/kg)
0	350
0.5	260
1	200
2	100
5	17

Use these data to determine the reaction order and the value of the rate constant.

Strategy. To determine the reaction order, a trial and error approach is often taken. From Table IV.3.A, if it is a zeroth-order reaction, the plot of concentration vs. time should be a straight line. The plot of ln(C) vs. time should be a straight line for first-order kinetics. If it is second order, the plot of (1/C) vs. time will be a straight line. The value of k is then obtained from the slope of the line.

Solution:

Many reactions of environmental concern are first-order reactions. First assume that it is first order and plot the concentration-time data on a semilog scale (Figure E. IV.3.1C).

A straight line fits the data very well, so the assumption of first-order kinetics is valid. The slope of the straight line can be determined as 0.263/hr. It should be noted that the rate constant in Eq. IV.2.3 is based on exponential with base e, and the plot in the figure is based on $\log_{10}$. Consequently, the value of k to be used in Eq. IV.2.3 should be the product of the slope from the semilog_{10} plot and 2.303 (which is the natural log of 10), that is,

$$k = (0.263)(2.303) = 0.606/\text{hr}.$$

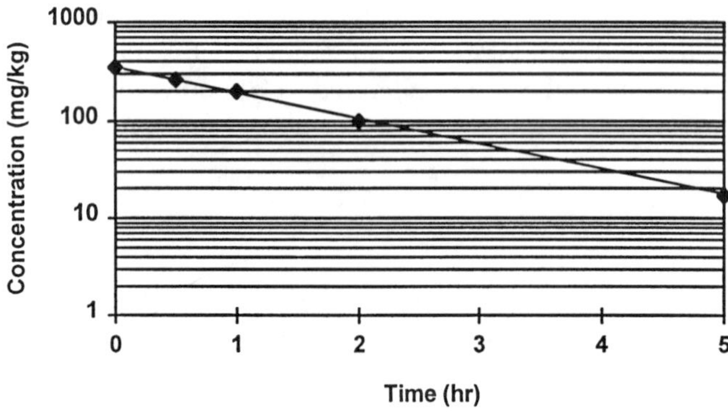

Figure E.IV.3.1C Concentration vs. time.

Discussion. Using the obtained rate constant and the initial concentration to calculate the concentration at other time t can serve as a check on the value. For example, the concentration, at t = 2 hours, can be calculated as (Eq. IV.3.2)

$$C_f = 350(e^{-(0.606)(2)}) = 104 \text{ mg/kg}$$

The calculated concentration, 104 mg/kg, is reasonably close to the reported experimental value, 100 mg/kg.

Example IV.3.1D Batch reactor with second-order kinetics

A batch reactor is to be designed to treat soil contaminated with 200 mg/kg of PCBs. The required removal, conversion, or reduction of PCBs is 90%. If the rate constant is $0.5[(\text{mg/kg})(\text{hr})]^{-1}$, what is the required residence time for the batch reactor?

Strategy. Although the order of the reaction is not mentioned in the problem statement, it is assumed to be a second-order reaction because the dimensions of k are $[(\text{mg/kg})(\text{hr})]^{-1}$.

Solution:

a. For a 90% reduction (η = 90%),

$$C_f = 200\ (1 - 90\%) = 20\ \text{mg/kg}$$

b. Insert the known values into Eq. IV.3.2:

$$20 = \frac{200}{1 + (0.5\tau)200}$$

$$\tau = 0.09\ \text{hr}$$

Discussion. The only difference between the reactors in Examples IV.3.1A and IV.3.1D is the reaction kinetics. With the same numerical value of the reaction rate constants, the required residence time to achieve the same conversation rate is much shorter in the reactor with second-order kinetics.

IV.3.2 CFSTRs

Let us now consider a steady-state CFSTR with a first order reaction. As mentioned earlier, by definition, the concentration in the effluent from a CFSTR is the same as that in the tank, and the concentration in the tank is uniform and constant. Under steady-state conditions, the flow rate is constant and $Q_{in} = Q_{out}$. By inserting Eq. IV.2.2 into Eq. IV.1.4, the mass balance equation can be expressed as:

$$\begin{aligned} 0 &= QC_{in} - QC_{out} + (V)(-kC_{reactor}) \\ &= QC_{in} - QC_{out} + (V)(-kC_{out}) \end{aligned} \qquad \text{[Eq. IV.3.5]}$$

With a simple mathematical manipulation, Eq. IV.3.5 can be rearranged as

$$\frac{C_{out}}{C_{in}} = \frac{1}{1+k(V/Q)} = \frac{1}{1+k\tau} \qquad \text{[Eq. IV.3.6]}$$

Table IV.3.B summarizes the design equations for CFSTRs in which zeroth-, first-, and second-order reactions take place.

Table IV.3.B Design Equations for CFSTRs

Reaction order, n	Design equation	
0	$C_{out} = C_{in} - k\tau$	[Eq. IV.3.7]
1	$\frac{C_{out}}{C_{in}} = \frac{1}{1+k\tau}$	same as [Eq. IV.3.6]
2	$\frac{C_{out}}{C_{in}} = \frac{1}{1+(k\tau)C_{out}}$	[Eq. IV.3.8]

Example IV.3.2A A soil slurry reactor with first-order kinetics (CFSTR)

A soil slurry reactor is used to treat soils contaminated with 1200 mg/kg of TPH. The required final soil TPH concentration is 50 mg/kg. From a bench-scale study, the rate equation is

$$\gamma = -0.25\ C \quad \text{in mg/kg/min}$$

The contents in the reactor are fully mixed. Assume that the reactor behaves as a CFSTR. Determine the required residence time to reduce the TPH concentration to 50 mg/kg.

Strategy. The format of the rate equation is a first-order reaction, and the reaction rate constant, k, is equal to 0.25/min.

Solution:

Insert the known values into Eq. IV.3.6 to find out the value of τ:

$$\frac{C_{out}}{C_{in}} = \frac{50}{1200} = \frac{1}{1+0.25\tau}$$

$$\tau = 92 \text{ minutes}$$

Example IV.3.2B *A low temperature heating soil reactor with second-order kinetics (CFSTR)*

A low temperature heating soil reactor is used to treat soil contaminated with 2500 mg/kg of TPH. The required final soil TPH concentration is 100 mg/kg. From a bench-scale study, the rate equation is

$$\gamma = -0.12\ C^2 \quad \text{in mg/kg/hr}$$

The reactor is rotated to achieve good mixing. Assume that the reactor behaves as a CFSTR. Determine the required residence time to reduce the TPH concentration to 100 mg/kg.

Strategy. The format of the rate equation is a second-order reaction, and the reaction rate constant, k, is equal to 0.12/(mg/kg)(hr).

Solution:

Insert the known values into Eq. IV.3.8, to find out the value of τ:

$$\frac{C_{out}}{C_{in}} = \frac{100}{1200} = \frac{1}{1+0.12\tau(100)}$$

$$\tau = 0.92 \text{ hours} = 55 \text{ minutes}$$

IV.3.3 PFRs

Let us now consider a steady-state PFR with a first-order reaction. As mentioned earlier, by definition, there is no longitudinal mixing within the PFR. The concentration in the reactor ($C_{reactor}$) decreases from C_{in} at the inlet point to C_{out} at the exit. Under the steady-state condition, the flow rate is constant and $Q_{in} = Q_{out}$. By inserting Eq. IV.2.2 into Eq. IV.1.4, the mass balance equation can be expressed as

$$0 = QC_{in} - QC_{out} + (V)(-kC_{reactor}) \qquad \text{[Eq. IV.3.9]}$$

The $C_{reactor}$ is a variable. The equation can be solved by considering an infinitesimal section of the reactor and integrating the equation. The solution can be expressed as:

$$\frac{C_{out}}{C_{in}} = e^{-k(V/Q)} = e^{-k\tau} \qquad \text{[Eq. IV.3.10]}$$

Table IV.3.C summarizes the design equations for PFRs in which zeroth-, first-, and second-order reactions take place.

Table IV.3.C Design Equations for PFRs

Reaction order, n	Design equation	
0	$C_{out} = C_{in} - k\tau$	[Eq. IV.3.11]
1	$C_{out} = C_{in}(e^{-k\tau})$	same as [Eq. IV.3.10]
2	$C_{out} = \dfrac{C_{in}}{1+(k\tau)C_{in}}$	[Eq. IV.3.12]

When comparing the design equations for PFRs in Table IV.3.C and for CFSTRs in Table IV.3.B, the following can be derived:

1. The zeroth-order reaction: The design equations are identical for both reactor types. It means that the conversion rate is independent of the reactor types, provided all the other conditions are the same.
2. The first-order reaction: The ratio of the outlet and inlet concentration is linearly proportional to the inverse of time for CFSTRs, and it is inversely and exponentially proportional to time for PFRs. In other words, the outlet concentration from PFRs decreases more sharply with increase of the residence time than that from CFSTRs, provided all the other conditions are the same. We can also say that, for a given residence time (or reactor size), the effluent concentration from a PFR would be lower than that from a CFSTR. (More discussions and examples will be given later in this section.)
3. Second-order reaction: The design equations for the second-order reactions are similar in format. The only difference is the C_{out} in the denominator on the right-hand side of Eq. IV.3.8 is replaced by C_{in} in Eq. IV.3.12. With a smaller value of C_{out} over C_{in}, the C_{out}/C_{in} ratio of a PFR will be smaller than that of a CFSTR. The smaller C_{out}/C_{in} ratio means that the effluent concentration would be lower for the same influent concentration.

Example IV.3.3A *A soil slurry reactor with first-order kinetics (PFR)*

A soil slurry reactor is used to treat soils contaminated with 1200 mg/kg of TPH. The required final soil TPH concentration is 50 mg/kg. From a bench-scale study, the rate equation is

$$\gamma = -0.25\ C \quad \text{in mg/kg/min}$$

Assume that the reactor behaves as a PFR. Determine the required residence time to reduce the TPH concentration to 50 mg/kg.

Strategy. The format of the rate equation is a first-order reaction, and the reaction rate constant, k, is equal to 0.25/min.

Solution:
Insert the known values into Eq. IV.3.10 to find the value of τ:

$$\frac{C_{out}}{C_{in}} = \frac{50}{1200} = e^{-(0.25)\tau}$$

$$\tau = 12.7 \text{ minutes}$$

Discussion. For the same inlet concentration and reaction rate constant, the required residence time to achieve a specified final concentration for a PFR, 12.7 minutes, is much smaller than that for a CFSTR, 92 minutes (see Example IV.3.2A).

For the first-order kinetics, the reaction rate is proportional to the concentration (i.e., $\gamma = KC_{reactor}$). The higher the reactor concentration, the higher the reaction rate. For CFSTRs, by definition, the reactor concentration is equal to the effluent concentration (i.e., 50 mg/kg in this case). For PFRs, by definition, the reactor concentration decreases from C_{in} (1200 mg/kg) at the inlet to C_{out} (50 mg/kg) at the outlet. The average concentration in the PFR (625 mg/kg as the arithmetic average or 245 as the geometric average) is much higher than 50 mg/kg, which makes the reaction rate much higher. Consequently, the required residence time would be much shorter.

Example IV.3.3B *A low temperature heating soil reactor with second-order kinetics (PFR)*

A low-temperature heating soil reactor is used to treat soils contaminated with 2500 mg/kg of TPH. The required final soil TPH concentration is 100 mg/kg. From a bench-scale study, the rate equation is

$$\gamma = -0.12\ C^2 \quad \text{in mg/kg/hr}$$

The soils are carried through the reactor on a conveyor belt. Assume that the reactor behaves as a PFR. Determine the required residence time to reduce the TPH concentration to 100 mg/kg.

Strategy. The format of the rate equation is a second-order reaction, and the reaction rate constant, k, is equal to 0.12/(mg/kg)(hr).

Solution:
Insert the known values into Eq. IV.3.12 to find the value of τ

$$\frac{C_{out}}{C_{in}} = \frac{100}{1200} = \frac{1}{1 + 0.12\tau(1200)}$$

τ = 0.08 hours = 4.8 minutes.

Discussion. Again, for the same inlet concentration and reaction rate constant, the required residence time to achieve a specified final concentration for a PFR, 4.8 minutes, is much smaller than that for a CFSTR, 55 minutes (see Example IV.3.2B).

IV.4 Sizing the reactors

Once the reactor type is selected and the required residence time to achieve the specified conversion is determined, sizing a reactor is straightforward. The longer the compound stays inside a reactor to achieve the desired conversion (i.e., the longer the residence time), the larger the reactor needed for a given flow rate.

For flow reactors such as CFSTRs and PFRs, the residence time, or the hydraulic detention time, τ, can be defined as

$$\tau = \frac{V}{Q} \qquad \text{[Eq. IV.4.1]}$$

where V is the volume of the reactor and Q is the flow rate. For a PFR, by definition, each fluid particle should spend exactly the same amount of time flowing through the reactor. On the other hand, for a CFSTR, most fluid particles would flow through the reactor in a shorter or longer time than the average retention time. Therefore, the value of τ in Eq. IV.4.1 is the average hydraulic retention time and is used in determining the size of the reactor.

For a batch reactor, the residence time calculated from Eqs. IV.3.2, IV.3.3, and IV.3.4 is the actual time needed for the reaction to be complete. To size the reactor, an engineer needs to include the time needed for loading, cool down, and unloading.

Example IV.4A Sizing a batch reactor

A soil slurry batch reactor is used to treat soils contaminated with 1200 mg/kg of TPH. It is necessary to treat the slurry at 30 gal/min. The required final soil TPH concentration is 50 mg/kg. From a bench-scale study, the rate equation is

$$\gamma = -0.05\,C \quad \text{in mg/kg/min}$$

The time required for loading and unloading the slurry for each batch is 2 hours. Size the batch reactor for this project.

Strategy. The format of the rate equation is a first-order reaction, and the reaction rate constant, *k*, is equal to 0.05/min.

Solution:

a. Insert the known values into Eq. IV.3.2 to find the value of τ:

$$\frac{C_{out}}{C_{in}} = \frac{50}{1200} = e^{-(0.05)\tau}$$

$$\tau = 64 \text{ min (needed for reaction)}$$

b. The total time needed for each batch = reaction time + time for loading and unloading = 64 + 120 = 184 minutes.
c. The required reactor volume, $V = (\tau)Q$ (from Eq. IV.4.1)

$$= (64 \text{ min})(30 \text{ gal/min}) = 1920 \text{ gal}$$

Discussion. A minimum of three reactors (1920 gallons each) are needed in this case. The reactors are operated in different phases; while two are in loading or unloading phases, the other one will be in active reaction phase. Consequently, the influent flow will not be interrupted.

Example IV.4B *Sizing a CFSTR*

A soil slurry reactor is used to treat soils contaminated with 1200 mg/kg of TPH. It is necessary to treat the slurry at 30 gal/min. The required final soil TPH concentration is 50 mg/kg. From a bench-scale study, the rate equation is

$$\gamma = -0.05\ C \quad \text{in mg/kg/min}$$

The contents in the reactor are fully mixed. Assume that the reactor behaves as a CFSTR. Size the CFSTR for this project.

Solution:

a. Insert the known values into Eq. IV.3.6 to find the value of τ:

$$\frac{C_{out}}{C_{in}} = \frac{50}{1200} = \frac{1}{1+(0.05)\tau}$$

$$\tau = 460 \text{ minutes}$$

b. The required reactor volume,

$$V = (\tau)Q \text{ (from Eq. IV.4.1)} = (460 \text{ min})(30 \text{ gal/min}) = 13{,}800 \text{ gal.}$$

Example IV.4C *Sizing a PFR*

A soil slurry reactor is used to treat soils contaminated with 1200 mg/kg of TPH. It is necessary to treat the slurry at 30 gal/min. The required final soil TPH concentration is 50 mg/kg. From a bench-scale study, the rate equation is

$$\gamma = -0.05\ C \quad \text{in mg/kg/min}$$

Assume that the reactor behaves as a PFR. Size the PFR for this project.

Solution:

a. Insert the known values into Eq. IV.3.10 to find the value of τ

$$\frac{C_{out}}{C_{in}} = \frac{50}{1200} = e^{-(0.05)\tau}$$

$$\tau = 64 \text{ minutes}$$

b. The required reactor volume, $V = (\tau)Q$ (from Eq. IV.4.1)

$$= (64 \text{ minutes})(30 \text{ gal/min}) = 1920 \text{ gal}$$

Discussion

1. To achieve the same conversion, the size of the PFR, 1920 gal (this example), is much smaller than 13,800 gal for the CFSTR (Example IV.4.B). The other advantage of PFRs is that all the influent flow receives the same residence time. This is extremely important for processes such as disinfection in a chlorine contact tank, in which all the fluid parcels should stay in the tank long enough to achieve the required kill. On the other hand, the complete mixing in the tank of the CFSTRs provides a great endurance to shock load. This is favorable for processes such as biological processes that are sensitive to shock load.
2. The design equations for batch reactors and PFRs are essentially the same. The required reaction times for these two reactors are the same, at 64 minutes. The actual tankage of the PFR is much smaller because loading and unloading need not be included in operation of flow reactors.

IV.5 Reactor configurations

In practical engineering applications, it is more common to have a few smaller reactors than to have one large reactor for the following reasons:

- Flexibility (to handle fluctuations of flow rate)
- Maintenance considerations
- A higher removal efficiency

Common reactor configurations include arrangement of reactors in series, in parallel, or a combination of both.

IV.5.1 Reactors in series

For reactors in series, the flow rates to all the reactors are the same and equal to the influent flow rate to the first reactor, Q (Figure IV.5A). The first reactor, with a volume V_1, will reduce the influent contaminant concentration, C_o, and yields an effluent concentration, C_1. The effluent concentration from the first reactor becomes the influent concentration to the second reactor. Consequently, the effluent concentration from the second reactor, C_2, becomes the influent concentration to the third reactor. More reactors can be added in series until the effluent concentration from the last reactor in the series meets the requirement. For CFSTRs, a few small reactors in series will yield a lower final effluent concentration than a large reactor with the same total volume. This will be illustrated by examples in this section.

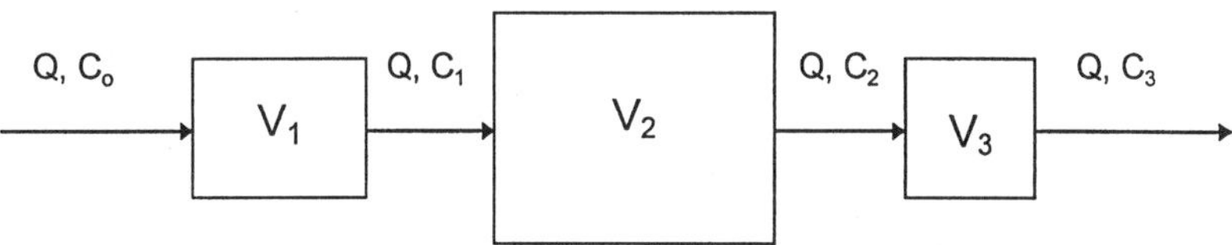

Figure IV.5A Three reactors in series.

For three CFSTRs arranged in series, the effluent concentration from the third reactor of CFSTRs in series can be determined from the contaminant concentration in the raw waste stream as

$$\frac{C_3}{C_o} = \left(\frac{C_3}{C_2}\right)\left(\frac{C_2}{C_1}\right)\left(\frac{C_1}{C_0}\right) = \left(\frac{1}{1+k_3\tau_3}\right)\left(\frac{1}{1+k_2\tau_2}\right)\left(\frac{1}{1+k_1\tau_1}\right) \quad \text{[Eq. IV.5.1]}$$

For three PFRs arranged in series, the effluent concentration from the third reactor of CFSTRs in series can be determined from the contaminant concentration in the raw waste stream as

$$\frac{C_3}{C_o} = \left(\frac{C_3}{C_2}\right)\left(\frac{C_2}{C_1}\right)\left(\frac{C_1}{C_0}\right) = (e^{-k_3\tau_3})(e^{-k_2\tau_2})(e^{-k_1\tau_1}) = e^{-(k_1\tau_1+k_2\tau_2+k_3\tau_3)} \quad \text{[Eq. IV.5.2]}$$

Example IV.5.1A CFSTRs in series

Subsurface soil at a site is contaminated with diesel fuel at a concentration of 1800 mg/kg. Above-ground remediation, using slurry bioreactors, is proposed. The treatment system is required to handle a slurry flow rate of 0.04 m^3/min. The required final diesel concentration in the soil is 100 mg/kg. The reaction is first-order with a rate constant 0.1/min, as determined from a bench-scale study.

Four different configurations of slurry bioreactors in the CFSTR mode are considered. Determine the final effluent concentration from each of these arrangements and if it meets the cleanup requirement:

a. One 4-m^3 reactor
b. Two 2-m^3 reactors in series
c. One 1-m^3 reactor followed by one 3-m^3 reactor
d. One 3-m^3 reactor followed by one 1-m^3 reactor

Solution:

a. For the 4-m^3 reactor, the residence time = V/Q = 4 m^3/(0.04 m^3/min) = 100 minutes.

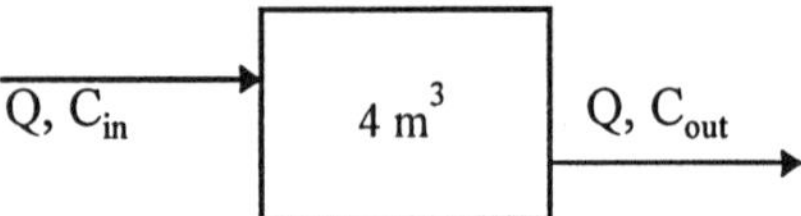

Use Eq. IV.3.6 to find the final effluent concentration

$$\frac{C_{out}}{C_{in}} = \frac{C_{out}}{1800} = \frac{1}{1+(0.1)(100)}$$

C_{out} = 164 mg/kg. (It exceeds the cleanup level.)

b. For the two 2-m^3 reactors, the residence time = V/Q = 2 m^3/(0.04 m^3/min) = 50 minutes each.

Q, C_o → 2 m^3 → Q, C_1 → 2 m^3 → Q, C_2

Use Eq. IV.5.1 to find the final effluent concentration

$$\frac{C_2}{C_o} = \left(\frac{C_2}{1800}\right) = \left(\frac{C_2}{C_1}\right)\left(\frac{C_1}{C_0}\right) = \left(\frac{1}{1+(0.1)(50)}\right)\left(\frac{1}{1+(0.1)(50)}\right)$$

C_{out} = 50 mg/kg. (It is below the cleanup level.)

c. The residence time of the first reactor = 1 m^3/(0.04 m^3/min) = 25 minutes. The residence time of the second reactor = 3 m^3/(0.04 m^3/min) = 75 minutes.

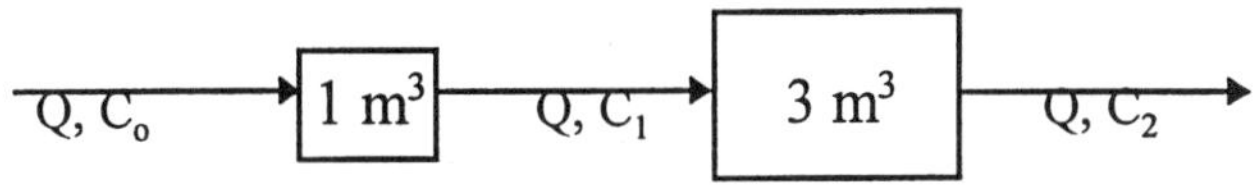

$$\xrightarrow{Q,\ C_0} \boxed{1\ m^3} \xrightarrow{Q,\ C_1} \boxed{3\ m^3} \xrightarrow{Q,\ C_2}$$

Use Eq. IV.5.1 to find the final effluent concentration

$$\frac{C_2}{C_o} = \left(\frac{C_2}{1800}\right) = \left(\frac{C_2}{C_1}\right)\left(\frac{C_1}{C_0}\right) = \left(\frac{1}{1+(0.1)(25)}\right)\left(\frac{1}{1+(0.1)(75)}\right)$$

C_{out} = 60.5 mg/kg. (It is below the cleanup level.)

d. The residence time of the first reactor = 3 m^3/(0.04 m^3/min) = 75 minutes. The residence time of the second reactor = 1 m^3/(0.04 m^3/min) = 25 minutes.

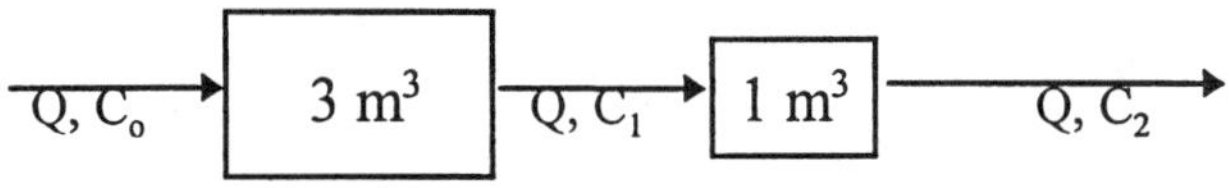

$$\xrightarrow{Q,\ C_0} \boxed{3\ m^3} \xrightarrow{Q,\ C_1} \boxed{1\ m^3} \xrightarrow{Q,\ C_2}$$

Use Eq. IV.5.1 to find the final effluent concentration

$$\frac{C_2}{C_o} = \left(\frac{C_2}{1800}\right) = \left(\frac{C_2}{C_1}\right)\left(\frac{C_1}{C_0}\right) = \left(\frac{1}{1+(0.1)(75)}\right)\left(\frac{1}{1+(0.1)(25)}\right)$$

C_{out} = 60.5 mg/kg. (It is below the cleanup level.)

Discussion

1. The total volume of the reactor(s) for each of the four configurations is 4 m^3.
2. The effluent concentration from the first set-up (one large reactor) is the highest. Actually, a series of smaller CFSTRs will always be more efficient in conversion than a single large CFSTR. A PFR can be viewed as an infinite series of small CFSTRs, and a PFR is always more efficient than a CFSTR of equal size.

3. For the configurations having two small reactors in series, the setup with two equal-size reactors yields the lowest effluent concentration.
4. For two reactors of different sizes, the sequence of the reactors does not affect the final effluent concentration, provided the rate constants in the reactors are the same.

Example IV.5.1B *PFRs in series*

Subsurface soil at a site is contaminated with diesel fuel at a concentration of 1800 mg/kg. Above-ground remediation, using slurry bioreactors, is proposed. The treatment system is required to handle a slurry flow rate of 0.04 m^3/min. The required final diesel concentration in the soil is 100 mg/kg. The reaction is first-order with a rate constant of 0.1/min, as determined from a bench-scale study.

Four different configurations of slurry bioreactors in the PFR mode are considered. Determine the final effluent concentration from each of these arrangements and if it meets the cleanup requirement:

a. One 4-m^3 reactor
b. Two 2-m^3 reactors in series
c. One 1-m^3 reactor followed by one 3-m^3 reactor
d. One 3-m^3 reactor followed by one 1-m^3 reactor

Solution:

a. For the 4-m^3 reactor, the residence time = V/Q = 4 m^3/(0.04 m^3/min) = 100 minutes.

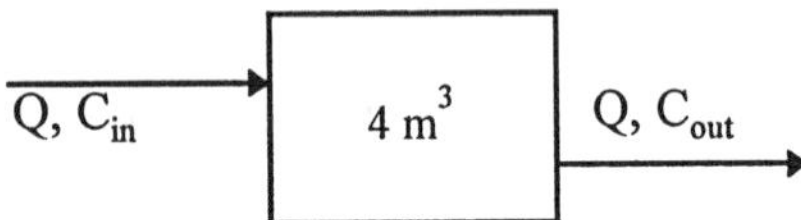

Use Eq. IV.3.10 to find the final effluent concentration

$$\frac{C_{out}}{C_{in}} = \frac{C_{out}}{1800} = e^{-(0.1)(100)}$$

$C_{out} = 8.2 \times 10^{-2}$ mg/kg. (It is below the cleanup level.)

b. For the two 2-m^3 reactors, the residence time = V/Q = 2 m^3/(0.04 m^3/min) = 50 minutes each.

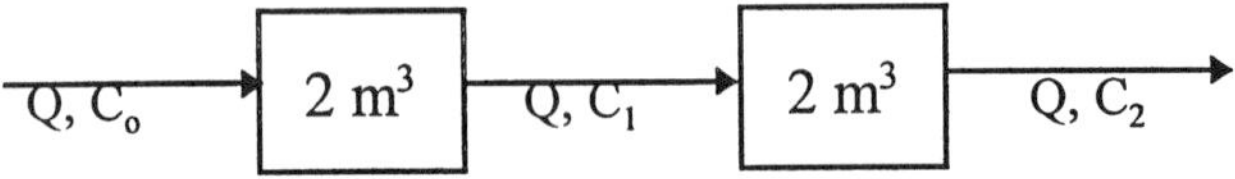

Use Eq. IV.5.2 to find the final effluent concentration

$$\frac{C_2}{C_o} = \left(\frac{C_2}{1800}\right) = \left(\frac{C_2}{C_1}\right)\left(\frac{C_1}{C_0}\right) = (e^{-(0.1)(50)})(e^{-(0.1)(50)}) = (e^{-(0.1)(50+50)})$$

$C_{out} = 8.2 \times 10^{-2}$ mg/kg. (It is below the cleanup level.)

c. The residence time of the first reactor = 1 m³/(0.04 m³/min) = 25 minutes. The residence time of the second reactor = 3 m³/(0.04 m³/min) = 75 minutes.

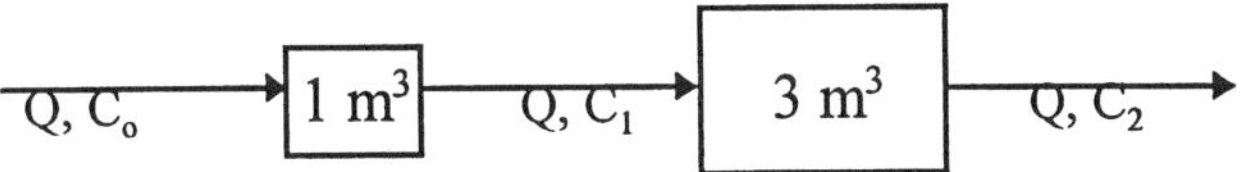

Use Eq. IV.5.1 to find the final effluent concentration

$$\frac{C_2}{C_o} = \left(\frac{C_2}{1800}\right) = \left(\frac{C_2}{C_1}\right)\left(\frac{C_1}{C_0}\right) = (e^{-(0.1)(25)})(e^{-(0.1)(75)}) = (e^{-(0.1)(25+75)})$$

$C_{out} = 8.2 \times 10^{-2}$ mg/kg. (It is below the cleanup level.)

d. The residence time of the first reactor = 3 m³/(0.04 m³/min) = 75 minutes. The residence time of the second reactor = 1 m³/(0.04 m³/min) = 25 minutes.

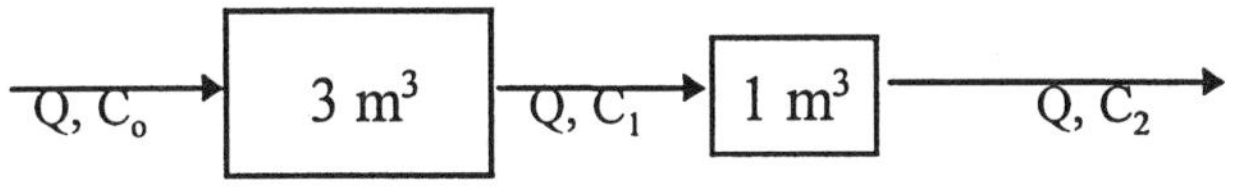

Use Eq. IV.5.1 to find the final effluent concentration

$$\frac{C_2}{C_o} = \left(\frac{C_2}{1800}\right) = \left(\frac{C_2}{C_1}\right)\left(\frac{C_1}{C_0}\right) = (e^{-(0.1)(75)})(e^{-(0.1)(25)}) = (e^{-(0.1)(75+25)})$$

$C_{out} = 8.2 \times 10^{-2}$ mg/kg. (It is below the cleanup level.)

Discussion

1. The total volume of the reactor(s) for each of the four configurations is 4 m³.

2. The effluent concentrations from all four different configurations are the same. It implies that the order of the reactors does not affect the final effluent concentration, provided the rate constants in the reactors and the total residence time of all the reactors in series are the same.
3. The effluent concentration of PFRs is much lower than those of CF-STRs in Example IV.5.1A.

Example IV.5.1C CFSTRs in series

Low-temperature-heating soil reactors (assuming they are ideal CFSTRs) are used to treat soils contaminated with 1050 mg/kg of TPH. The required final soil TPH concentration is 10 mg/kg. A reactor with a 20-minute residence time can only reduce the concentration to 50 mg/kg. Assume the reaction is first-order. Can two smaller reactors (10 minutes residence time each) in series reduce the TPH concentration below 10 mg/kg?

Strategy. The reaction rate constant was not given, so we have to find the value first.

Solution:

a. Use Eq. IV.3.6 to find the rate constant:

$$\frac{C_{out}}{C_{in}} = \frac{50}{1050} = \frac{1}{1+(k)(20)}$$

$$k = 1/\text{min}$$

b. For the two small reactors in series:
Use Eq. IV.5.1 to find out the final effluent concentration

$$\frac{C_2}{C_o} = \left(\frac{C_2}{1050}\right) = \left(\frac{C_2}{C_1}\right)\left(\frac{C_1}{C_0}\right) = \left(\frac{1}{1+(1)(10)}\right)\left(\frac{1}{1+(1)(10)}\right)$$

C_{out} = 8.7 mg/kg. (It is below the cleanup level.)

Discussion. This example again demonstrates that two smaller CFSTRs can do a better job than a larger CFSTR with an equivalent total volume. However, two reactors may require a larger capital investment (two sets of process controls for example) and higher O&M costs.

Example IV.5.1D PFRs in series

UV/ozone treatment is selected to remove TCE from a recovered groundwater stream (TCE concentration = 200 ppb). At a design flow rate of 50

L/min an off-the-shelf reactor would provide a hydraulic retention of 5 minutes and reduce TCE concentration from 200 ppb to 16 ppb. However, the discharge limit for TCE is 3.2 ppb. Assuming the reactors are of ideal plug flow type and the reaction is first-order, how many reactors would you recommend? What would be the TCE concentration in the final effluent?

Solution:

a. Use Eq. IV.3.10 to find out the reaction rate constant

$$\frac{C_{out}}{C_{in}} = \frac{16}{200} = e^{-(k)(5)}$$

$$k = 0.505/\text{min.}$$

b. Use Eq. IV.5.2 to find out the final effluent concentration from two PFRs in series

$$\frac{C_2}{C_o} = \left(\frac{C_2}{200}\right) = \left(\frac{C_2}{C_1}\right)\left(\frac{C_1}{C_0}\right) = (e^{-(0.505)(5)})(e^{-(0.505)(5)})$$

C_{out} = 1.28 ppb. (It is less than 3.2 ppb.)

Two PFRs, each with 5 minutes residence time, would be needed.

Discussion. We can also determine the total residence time needed to reduce the final concentration to 3.2 ppb first and then determine the number of PFRs needed. Use Eq. IV.3.10 to find out the required retention time.

$$\frac{C_{out}}{C_{in}} = \frac{3.2}{200} = e^{-(0.505)(\tau)}$$

τ = 8.2 minutes. (Two PFRs needed.)

IV.5.2 Reactors in parallel

For reactors in parallel, the reactors share the same influent (the influent is split and fed to the reactors). The flow rate to each reactor in parallel can be different; however, the influent concentration to all the reactors in parallel should be identical. The sizes of the reactors may not be the same, and the effluent concentrations from the reactors can be different (Figure IV.5.B). In that figure, the following mass balance equations are valid:

$$Q = Q_1 + Q_2 \qquad \text{[IV.5.3]}$$

$$C_f = \frac{Q_1C_1 + Q_2C_2}{Q_1 + Q_2} \qquad \text{[IV.5.4]}$$

Reactors in-parallel configurations are often used for the following cases (1) a single reactor cannot handle the flow rate, (2) the total influent rate fluctuates significantly, or (3) the reactors require frequent maintenance.

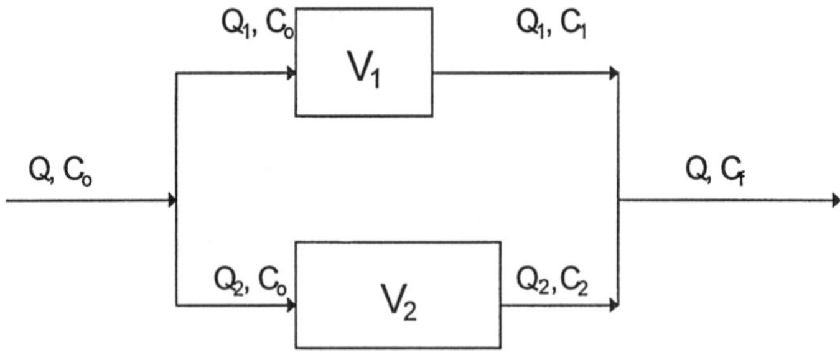

Figure IV.5.B Two reactors in parallel.

Example IV.5.2A CFSTRs in parallel

Subsurface soil at a site is contaminated with diesel fuel at a concentration of 1800 mg/kg. Above-ground remediation, using slurry bioreactors, is proposed. The treatment system is required to handle a slurry flow rate of 0.04 m^3/min. The required final diesel concentration in the soil is 100 mg/kg. The reaction is first-order with a rate constant of 0.1/min, as determined from a bench-scale study.

Four different configurations of slurry bioreactors in the CFSTR mode are considered. Determine the final effluent concentration from each of these arrangements and if it meets the cleanup requirement:

a. One 4-m^3 reactor
b. Two 2-m^3 reactors in parallel (each receives 0.02 m^3/min flow)
c. One 1-m^3 reactor and one 3-m^3 reactor (each receives 0.02 m^3/min flow) in parallel
d. One 1-m^3 reactor and one 3-m^3 reactor (the smaller one receives 0.01 m^3/min flow and the other receives 0.03 m^3/min) in parallel

Solution:

a. For the 4-m^3 reactor, the residence time = V/Q = 4 m^3/(0.04 m^3/min) = 100 minutes.

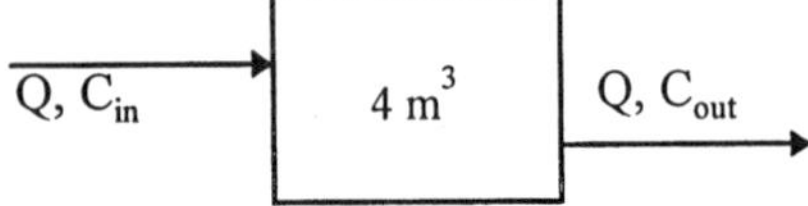

Use Eq. IV.3.6 to find the final effluent concentration

$$\frac{C_{out}}{C_{in}} = \frac{C_{out}}{1800} = \frac{1}{1+(0.1)(100)}$$

C_{out} = 164 mg/kg. (It exceeds the cleanup level.)

b. For the two 2-m^3 reactors, the residence time = V/Q = 2 m^3/(0.02 m^3/min) = 100 minutes each.

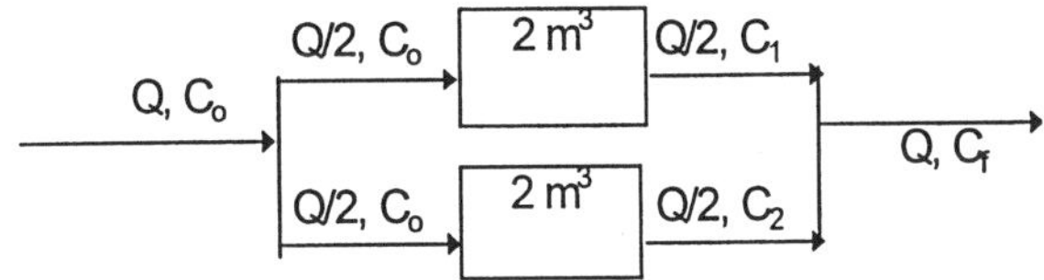

Use Eq. IV.3.6 to find the effluent concentration from each reactor

$$\frac{C_{out}}{C_{in}} = \frac{C_{out}}{1800} = \frac{1}{1+(0.1)(100)}$$

C_{out} = 164 mg/kg. for both reactors and the combined final effluent. It exceeds the cleanup level.

c. The residence time of the first reactor = 1 m^3/(0.02 m^3/min) = 50 minutes. The residence time of the second reactor = 3 m^3/(0.02 m^3/min) = 150 minutes.

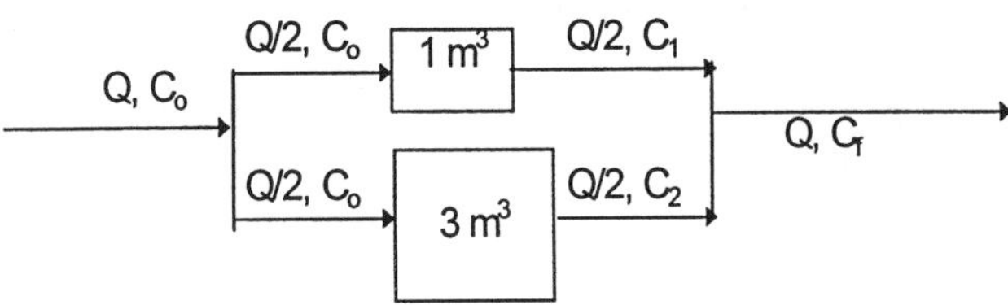

Use Eq. IV.3.6 to find the effluent concentration from each reactor. Reactor 1:

$$\frac{C_1}{1800} = \frac{1}{1+(0.1)(50)}$$

$$C_1 = 300 \text{ mg/kg}$$

Reactor 2:

$$\frac{C_1}{1800} = \frac{1}{1+(0.1)(150)}$$

$$C_2 = 112.5 \text{ mg/kg}$$

Use Eq. IV.5.4 to find the concentration of the combined effluent

$$C_f = \frac{2(300)+2(112.5)}{2+2} = 206 \text{ mg/kg}$$

It exceeds the cleanup level.

d. The residence time of the first reactor = 1 m³/(0.01 m³/min) = 100 minutes. The residence time of the second reactors = 3 m³/(0.03 m³/min) = 100 minutes.

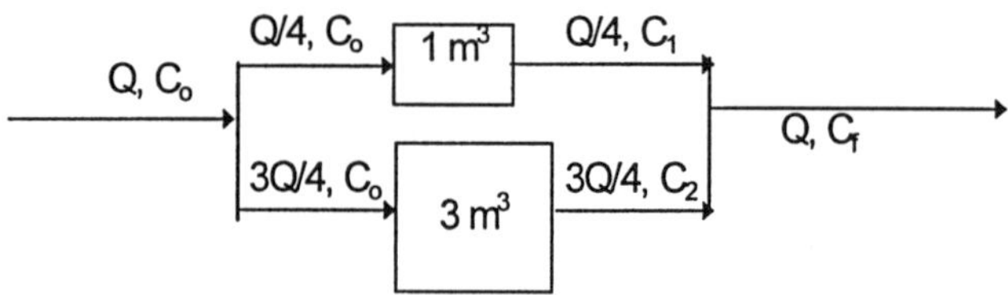

Use Eq. IV.3.6 to find the effluent concentration from each reactor

$$\frac{C_{out}}{C_{in}} = \frac{C_{out}}{1800} = \frac{1}{1+(0.1)(100)}$$

C_{out} = 164 mg/kg for both reactors and the combined final effluent. It exceeds the cleanup level.

Discussion

1. The total volume of the reactor(s) for each of the four configurations is 4 m³.
2. The effluent concentrations from all the configurations exceed the cleanup level. The configurations (a), (b), and (d) have the same effluent concentrations because the residence times of all reactors are identical. The effluent concentration from configuration (c) is the worst among the four.
3. To split the flow into reactors in parallel with the same residence time does not have any impact on the final effluent concentration, i.e., case (a) and case (b).

Example IV.5.2B *PFRs in parallel*

Subsurface soil at a site is contaminated with diesel fuel at a concentration of 1800 mg/kg. Above-ground remediation, using slurry bioreactors, is proposed. The treatment system is required to handle a slurry flow rate of 0.04 m^3/min. The required final diesel concentration in the soil is 100 mg/kg. The reaction is first-order with a rate constant of 0.1/min, as determined from a bench-scale study.

Four different configurations of slurry bioreactors in the PFR mode are considered. Determine the final effluent concentration from each of these arrangements and if it meets the cleanup requirement:

a. One 4-m^3 reactor
b. Two 2-m^3 reactors in parallel (each receives 0.02 m^3/min flow)
c. One 1-m^3 reactor and one 3-m^3 reactor (each receives 0.02 m^3/min flow) in parallel
d. One 1-m^3 reactor and one 3-m^3 reactor (the smaller one receives 0.01 m^3/min flow and the other receives 0.03 m^3/min) in parallel

Solution:

a. For the 4-m^3 reactor, the residence time = V/Q = 4 m^3/(0.04 m^3/min) = 100 minutes.

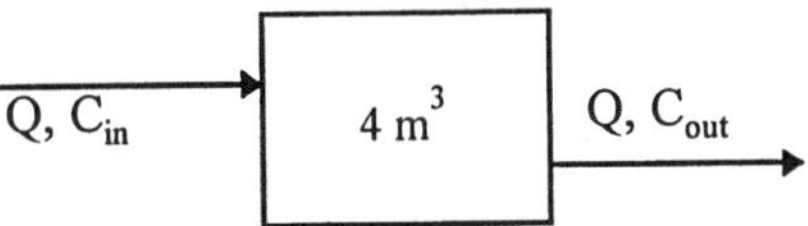

Use Eq. IV.3.10 to find the final effluent concentration

$$\frac{C_{out}}{C_{in}} = \frac{C_{out}}{1800} = e^{-(0.1)(100)}$$

C_{out} = 8.2 × 10^{-2} mg/kg. (It is below the cleanup level.)

b. For the two 2-m^3 reactors, the residence time = V/Q = 2 m^3/(0.02 m^3/min) = 100 minutes each.

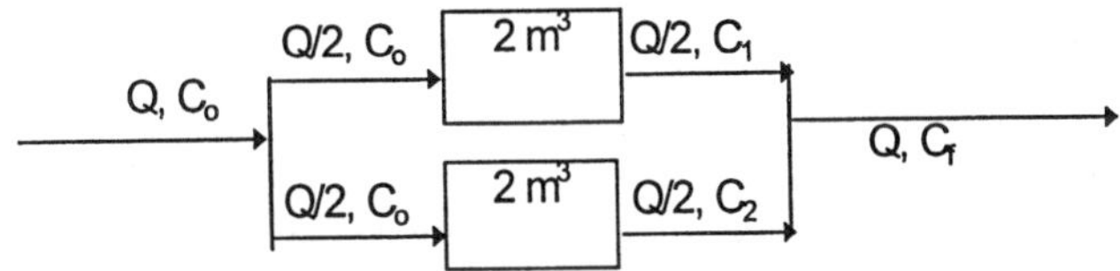

Use Eq. IV.3.10 to find the effluent concentration for each reactor

$$\frac{C_{out}}{C_{in}} = \frac{C_{out}}{1800} = e^{-(0.1)(100)}$$

$C_{out} = 8.2 \times 10^{-2}$ mg/kg for each reactor and the combined effluent. It is below the cleanup level.

c. The residence time of the first reactor = 1 m^3/(0.02 m^3/min) = 50 minutes. The residence time of the second reactor = 3 m^3/(0.02 m^3/min) = 150 minutes.

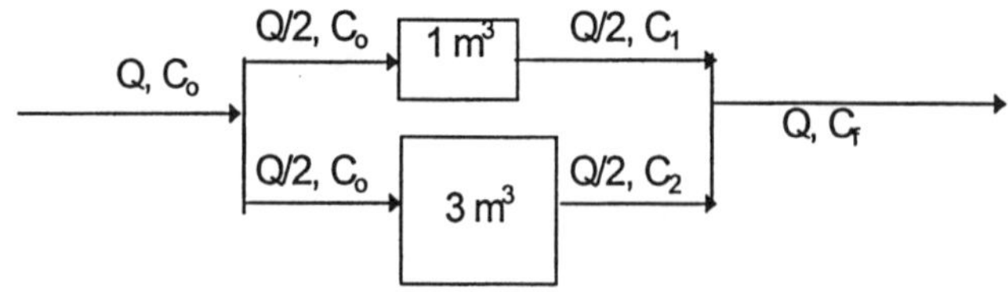

Use Eq. IV.3.6 to find the effluent concentration from each reactor. Reactor 1:

$$\frac{C_{out}}{1800} = e^{-(0.1)(50)}$$

$$C_1 = 12.2 \text{ mg/kg.}$$

Reactor 2:

$$\frac{C_{out}}{1800} = e^{-(0.1)(150)}$$

$$C_2 = 5.5 \times 10^{-4} \text{ mg/kg.}$$

Use Eq. IV.5.4 to find out the concentration of the combined effluent

$$C_f = \frac{2(12.2) + 2(5.5 \times 10^{-4})}{2+2} = 6.1 \text{ mg/kg}$$

It is below the cleanup level.

d. The residence time of the first reactor = 1 m^3/(0.01 m^3/min) = 100 minutes. The residence time of the second reactor = 3 m^3/(0.03 m^3/min) = 100 minutes.

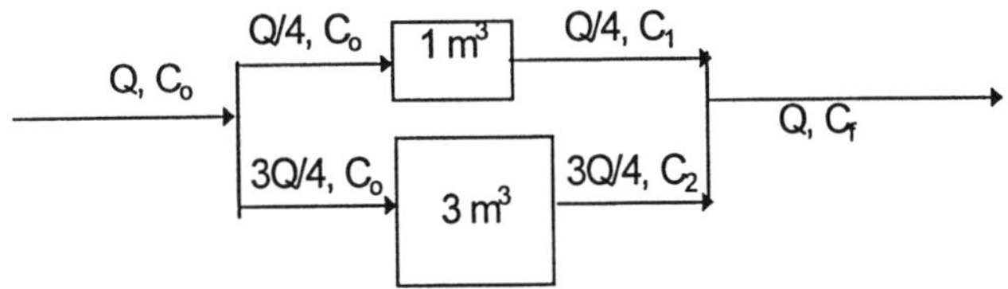

Use Eq. IV.3.10 to find the effluent concentration from each reactor:

$$\frac{C_{out}}{C_{in}} = \frac{C_{out}}{1800} = e^{-(0.1)(100)}$$

C_{out} = 8.2 × 10^{-2} mg/kg for both reactors and the combined final effluent. It is below the cleanup level.

Discussion

1. The total volume of the reactor(s) for each of the four configurations is 4 m^3.
2. The effluent concentrations from all the configurations are below the cleanup level. The configurations (a), (b), and (d) have the same effluent concentrations because the residence times of all reactors are identical. The effluent concentration from configuration (c) is the worst among the four.
3. To split the flow into reactors in parallel with the same residence time does not have any impact on the final effluent concentration.

chapter five

Vadose zone soil remediation

This chapter illustrates important design calculations for commonly used in situ and above-ground soil remediation techniques. The treatment processes covered include soil vapor extraction, soil bioremediation, soil washing, and low-temperature heating.

V.1 Soil vapor extraction

V.1.1 Introduction

Description of the soil venting process

Soil vapor extraction (SVE), also known as soil venting, in situ vacuum extraction, in situ volatilization, or soil vapor stripping, has become a very popular remediation technique for soil contaminated with VOCs. The process strips volatile organic constituents from contaminated soil by inducing an air flow through the contaminated zone. The air flow is created by a vacuum pump (often called a "blower") through a single well or network of wells.

As the soil vapor is swept away from the voids of the vadose zone, fresh air is naturally (through passive venting wells or air infiltration) or mechanically (through air injection wells) introduced and refills the voids. This flux of the fresh air will (1) disrupt the existing partition of the contaminants among the void, soil moisture, and soil grain surface by promoting volatilization of the adsorbed and dissolved phase of contaminants, (2) provide oxygen to indigenous microorganisms for biodegradation of the contaminants, and (3) carry away the toxic metabolic by-products generated from the biodegradation process. The extracted air is usually laden with VOCs and brought to the ground surface by the vacuum blower. Treatment of the extracted vapor is normally required. Design calculations for the VOC-laden air treatment are covered in Chapter seven.

Major components of an SVE system

Major components of a typical soil venting system include vapor extraction well(s), vacuum blower(s), moisture removal device (knock-out drum), off-gas collection piping and ancillary equipment, and the off-gas treatment system.

Important design considerations

The most important parameters for preliminary design are the extracted VOC concentration, air flow rate, radius of influence of the venting well, number of wells required, and size of the vacuum blower.

V.1.2 Expected vapor concentration

As mentioned in Section II.3.5, volatile organic contaminants in a vadose zone may be present in four phases: (1) in the soil moisture due to dissolution, (2) on the soil grain surface due to adsorption, (3) in the pore void due to volatilization, and (4) as the free product. If the free-product phase is present, the vapor concentration in the pore void can be estimated from Raoult's law as

$$P_A = (P^{vap})(x_A) \qquad \text{[Eq. V.1.1]}$$

where P_A = partial pressure of compound A in the vapor phase, P^{vap} = vapor pressure of compound A as a pure liquid, and x_A = mole fraction of compound A in the liquid phase.

Examples using Raoult's law can be found in Section II.3. The partial pressure calculated from Eq. V.1.1 represents the upper limit of the contaminant concentration in the extracted vapor from a soil venting project. The actual concentration will be lower than this upper limit because (1) not all the extracted air passes through the contaminated zone and (2) limitations on mass transfer exist. Nevertheless, this concentration serves as a starting point for estimating the initial vapor concentration at the beginning of a venting project. Initially the extracted vapor concentrations will be relatively constant. As soil venting continues, the free product phase will disappear. The extracted vapor concentration will then begin to drop, and the extracted vapor concentration will become dependent on the partitioning of the contaminants among the three other phases. As the air flows through the pores and sweeps away the contaminants, the contaminants dissolved in the soil moisture will volatilize from the liquid into the void. Simultaneously, the contaminants will also desorb from the soil grain surface and enter into the soil moisture (assuming the soil grains are covered by a moisture layer). Thus, the concentrations in all three phases decrease as the venting process progresses.

Table V.1.A Physical Properties of Gasoline and Weathered Gasoline

Compound	Molecular weight (g/mole)	P^{vap} @ 20°C (atm)	G_{est} ppmV	G_{est} mg/L
Gasoline	95	0.34	340,000	1343
Weathered gasoline	111	0.049	49,000	220

Modified from Johnson, P. C., Stanley, C. C., Kemblowski, M. W., Byers, D. L., and Colthart, J. D., *Ground Water Monitor. Rev.*, Spring, 1990b. With permission.

The above phenomenon describes common observations at sites that contain a single type of contaminant. Soil venting has also been widely used for sites contaminated with a mixture of compounds, such as gasoline. For these cases, the vapor concentration decreases continuously from the start of venting; a period of constant vapor concentration in the beginning phase of the project does not exist. This can be explained by the fact that each compound in the mixture has a different vapor pressure. Thus, the more volatile compounds tend to leave the free product, as well as the moisture and the soil surface, earlier and be extracted earlier. Table V.1.A shows the molecular weights of fresh and weathered gasoline and their vapor pressures at 20°C. The table also lists the saturated vapor concentrations that are in equilibrium with the fresh and weathered gasoline.

To estimate the initial concentration of the extracted vapor in equilibrium with the free-product phase, the following procedure can be used:

Step 1: Obtain the vapor pressure data of the compound of concern (e.g., from Table II.3.C).

Step 2: Determine the mole fraction of the compound in the free product. For a pure compound, set $x_A = 1$. For a mixture, follow the procedure in Section II.1.4.

Step 3: Use Eq. V.1.1 to determine the vapor concentration in atm or mmHg unit.

Step 4: Convert the concentration by volume into a mass concentration, if needed, by using Eq. II.1.1.

Information needed for this calculation

- Vapor pressure of the contaminant
- Molecular weight of the compounds

Example V.1.2A *Estimate the saturated gasoline vapor concentration*

Use the information in Table V.1.A to estimate the maximum gasoline vapor concentration from two soil venting projects. Both sites are contaminated from accidental gasoline spills. The spill at the first site happened recently, while the spill at the other site occurred 3 years ago.

Solution:

a. The site with fresh gasoline. Vapor pressure of fresh gasoline is 0.34 atm at 20°C, as shown in Table V.1.A. The partial pressure of this gasoline in the pore space can be found by using Eq. V.1.1. as:

$$P_A = (P^{vap})(x_A) = (0.34 \text{ atm})(1.0) = 0.34 \text{ atm}$$

Thus, the partial pressure of gasoline in the air is 0.34 atm (= 340,000 × 10^{-6} atm), which is equivalent to 340,000 ppmV. Use Eq. II.1.1 to convert the ppmV concentration into a mass concentration unit (at 20°C), as

$$1 \text{ ppmV fresh gasoline} = \{(\text{MW of fresh gasoline})/24.05\} \text{ mg/m}^3$$
$$= (95)/24.05 = 3.95 \text{ mg/m}^3$$

So,

$$340{,}000 \text{ ppmV} = (340{,}000)(3.95) = 1{,}343{,}000 \text{ mg/m}^3 = 1343 \text{ mg/L}$$

b. The site with weathered gasoline. Vapor pressure (as well as the partial pressure in this case) of weathered gasoline is 0.049 atm, which is equivalent to 49,000 ppmV. Use Eq. II.1.1 to convert the ppmV concentration into a mass concentration unit (at 20°C), as

$$1 \text{ ppmV weathered gasoline}$$
$$= \{(\text{MW of weathered gasoline})/24.05\} \text{ mg/m}^3$$
$$= (111)/24.05 = 4.62 \text{ mg/m}^3$$

So,

$$49{,}000 \text{ ppmV} = (49{,}000)(4.62) = 226{,}000 \text{ mg/m}^3 = 226 \text{ mg/L}$$

Discussion

1. The saturated vapor concentration of the weathered gasoline can be a few times less than that of the fresh gasoline. (In this case, it is more than five times smaller.)
2. The calculated vapor concentrations are essentially the same as those listed in Table V.1.A.
3. Although gasoline is a mixture of compounds, the mole fraction was set to one since the vapor pressure and molecular weight of gasoline were given as the weighted averages.

Example V.1.2B *Estimate saturated vapor concentrations of a binary mixture*

A site is contaminated with an industrial solvent. The solvent consists of 50% toluene and 50% xylenes by weight. Soil venting is considered for site remediation. Estimate the maximum vapor concentration of the extracted vapor. The subsurface temperature of the site is 20°C.

Solution:

a. From Table II.3.C, the following physicochemical properties were obtained: molecular weight = 92.1 (toluene) = 106.2 (xylenes) and P^{vap} = 22 mmHg (toluene) = 10 mmHg (xylenes).
b. The mole fraction of toluene in the solvent can be found as

$$\text{basis} = 1000 \text{ g solvent}$$

$$\text{moles of toluene} = \text{mass/MW} = [(50\%)(1000)] \div (92.1) = 5.43 \text{ moles}$$

$$\text{moles of xylenes} = \text{mass/MW} = [(50\%)(1000)] \div (106.2) = 4.71 \text{ moles}$$

$$\text{mole fraction of toluene} = (5.43)/(5.43 + 4.71) = 0.536$$

$$\text{mole fraction of xylenes} = 1 - 0.536 = 0.464$$

c. The saturated vapor concentration can be found by using Eq. V.1.1 as

$$P_{toluene} = (P^{vap})(x_A) = (22 \text{ mmHg})(0.536) = 11.79 \text{ mmHg} = 0.0155 \text{ atm}$$

Thus, partial pressure of toluene = 0.0155 atm = 15,500 ppmV

$$P_{xylenes} = (P^{vap})(x_A) = (10 \text{ mmHg})(0.464) = 4.64 \text{ mmHg} = 0.0061 \text{ atm}$$

Thus, partial pressure of xylenes = 0.0061 atm = 6,100 ppmV.

The volumetric (or molar) composition of the extracted vapor = (15,500)/[15,500 + 6100] = 71.8% ← toluene.

d. The mass concentration can be found by using Eq. II.1.1 as

$$1 \text{ ppmV toluene} = (92.1)/24.05 = 3.83 \text{ mg/m}^3$$

So,

$$15{,}500 \text{ ppmV} = (15{,}500)(3.83) = 59{,}400 \text{ mg/m}^3 = 59.4 \text{ mg/L}$$

$$1 \text{ ppmV xylenes} = (106.2)/24.05 = 4.42 \text{ mg/m}^3$$

So,

$$6100 \text{ ppmV} = (6100)(4.42) = 27{,}000 \text{ mg/m}^3 = 27.0 \text{ mg/L}$$

The weight composition of the extracted vapor = (59.4)/[59.4 +27.0] = 68.8% ← toluene.

Discussion

1. The toluene concentration in the extracted vapor is 68.8% by weight, that is higher than its concentration in the liquid solvent, 50% by weight. The higher percentage of toluene in the vapor is due to its higher vapor pressure.
2. This saturated vapor concentration would be higher than the actual concentration of the extracted vapor due to the fact (1) not all the air flows through the contaminated zone and (2) limitations on mass transfer exist.

As mentioned, the presence or absence of a free-product phase greatly affects the extracted vapor concentration. To determine if the free-product phase is present, the following procedure can be used:

Step 1: Obtain the physicochemical data of the compound of concern (e.g., from Table II.3.C).
Step 2: Assume the free-product phase is present. Use Eq. V.1.1 to determine the saturated vapor concentration in atm or mmHg unit.
Step 3: Convert the saturated vapor concentration into a mass concentration by using Eq. II.1.1.
Step 4: Determine the K_{oc} value using Eq. II.3.14 and determine the K_p value using Eq. II.3.12.
Step 5: Determine the contaminant concentration in soil by using Eq. II.3.23 and the vapor concentration from Step 3.
Step 6: If the contaminant concentration in soil determined from Step 5 is lower than the concentrations of the soil samples, the free-product phase should be present.

Information needed for this calculation

- Vapor pressure of the contaminant
- Molecular weight of the compound
- Henry's constant of the compound
- Organic water partition coefficient, K_{ow}
- Organic content, f_{oc}
- Porosity, ϕ
- Degree of water saturation
- Bulk density of soil, ρ_b

Example V.1.2C *Determine if the free-product phase is present in the subsurface*

A subsurface is contaminated by a spill of 1,1,1-trichloroethane (1,1,1-TCA). The TCA concentrations of the soil samples from the contaminated zone were between 5000 and 9000 mg/kg. The subsurface has the following characteristics:

Porosity = 0.4
Organic content in soil = 0.02
Degree of water saturation = 30%
Subsurface temperature = 20°C
Bulk density of soil = 1.8 g/cm^3

Determine if the free-product phase of TCA is present in the subsurface. What could be the maximum contaminant concentration in soil if the free-product phase of TCA is absent?

Solution:

a. From Table II.3.C, the following physicochemical properties of 1,1,1-TCA were obtained: Molecular weight = 133.4, H = 14.4 atm/M, P^{vap} = 100 mmHg, and Log K_{ow} = 2.49.
b. Use Eq. V.1.1 to determine the saturated TCA vapor concentration:

$$P^{vap} = 100 \text{ mmHg} = 0.132 \text{ atm} = 132{,}000 \text{ ppmV}$$

c. Convert the saturated vapor concentration into a mass concentration by using Eq. II.1.1:

$$1 \text{ ppmV TCA} = (133.4)/24.05 = 5.55 \text{ mg/m}^3$$

So,

$$G = 132{,}000 \text{ ppmV} = (132{,}000)(5.55) = 733{,}000 \text{ mg/m}^3 = 733 \text{ mg/L}$$

d. Use Table II.3.B to convert the Henry's constant to a dimensionless value:

$$H^* = H/RT = (14.4)/[(0.082)(273 + 20)] = 0.60 \text{ (dimensionless)}$$

Use Eq. II.3.14 to find K_{oc},

$$K_{oc} = 0.63K_{ow} = 0.63\,(10^{2.49}) = (0.63)(309) = 195$$

Use Eq. II.3.12 to find K_p,

$$K_p = f_{oc}K_{oc} = (0.02)\,(195) = 3.9 \text{ L/kg}$$

e. Use Eq. II.3.23 to estimate the soil concentration of TCA:

$$\frac{M_t}{V} = \left[\frac{(\phi_w)}{H} + \frac{(\rho_b)K_p}{H} + (\phi_a)\right]G$$

$$= \left[\frac{(0.4)(30\%)}{0.60} + \frac{(1.8)(3.9)}{0.60} + (0.4)(1-30\%)\right](733) = 8930 \text{ mg/L}$$

Divide the value by the bulk density of the soil to express the soil concentration in mg/kg:

Soil concentration = 8930 mg/L ÷ 1.8 kg/L = 4960 mg/kg

This value, 4960 mg/kg, represents the maximum contaminant concentration in the soil if the free-product phase of TCA is absent.

f. Since the calculated TCA concentration, 4960 mg/kg, is less than those of the soil samples, the free product phase of TCA should be present in the subsurface.

To determine the extracted soil vapor concentration in the absence of free-product in the subsurface, the following procedure can be used:

Step 1: Obtain the physicochemical data of the compound of concern (e.g., from Table II.3.C).
Step 2: Determine the K_{oc} value using Eq. II.3.14 and determine the K_p value using Eq. II.3.12.
Step 3: Convert the contaminant concentration in soil from mg/kg to mg/L.
Step 4: Determine the vapor concentration by using Eq. II.3.23 and the contaminant concentration in soil from Step 3.

Information needed for this calculation

- Contaminant concentrations of soil samples
- Henry's constant of the compound
- Organic water partition coefficient, K_{ow}
- Organic content, f_{oc}
- Porosity, ϕ
- Degree of water saturation
- Bulk density of soil, ρ_b

Example V.1.2D Estimate the extracted vapor concentration (in the absence of the free-product phase)

A subsurface is contaminated by a benzene spill. The average benzene concentration of the soil samples, taken from the contaminated zone, was 500 mg/kg. The subsurface has the following characteristics:

Porosity = 0.35
Organic content = 0.03
Water saturation = 45%
Subsurface temperature = 25°C
Bulk density of soil = 1.7 g/cm^3

Estimate the extracted soil vapor concentration at the start of the soil venting project.

Solution:

a. From Table II.3.C, the following physicochemical properties of benzene were obtained: molecular weight = 78.1, H = 5.55 atm/M, P^{vap} = 95.2 mmHg, and Log K_{ow} = 2.13.

b. Use Table II.3.B to convert the Henry's constant to a dimensionless value:

$$H^* = H/RT = (5.55)/[(0.082)(273 + 25)] = 0.23 \text{ (dimensionless)}$$

Use Eq. II.3.14 to find K_{oc},

$$K_{oc} = 0.63K_{ow} = 0.63\ (10^{2.13}) = (0.63)(135) = 85$$

Use Eq. II.3.12 to find K_p,

$$K_p = f_{oc}K_{oc} = (0.03)(85) = 2.6 \text{ L/kg}$$

c. Multiply the concentrations of the soil samples by the bulk density of the soil to express the soil concentration in mg/L:

$$\text{Soil concentration} = 500 \text{ mg/kg} \times 1.7 \text{ kg/L} = 850 \text{ mg/L}$$

d. Use Eq. II.3.23 to estimate the soil vapor concentration of benzene in equilibrium with this contaminant concentration in soil:

$$\frac{M_t}{V} = \left[\frac{(\phi_w)}{H} + \frac{(\rho_b)K_p}{H} + (\phi_a)\right]G$$

$$= \left[\frac{(0.35)(45\%)}{0.23} + \frac{(1.7)(2.6)}{0.23} + (0.35)(1-45\%)\right]G = 850 \text{ mg/L}$$

So,

$$G = 42.3 \text{ mg/L} = 42{,}300 \text{ mg/m}^3$$

e. Convert the vapor concentration into a volume concentration by using Eq. II.1.1:

$$1 \text{ ppmV benzene} = (78.1)/24.5 = 3.2 \text{ mg/m}^3 \text{ @ } 25°\text{C}$$

$$42{,}300 \text{ mg/m}^3 = 42{,}300 \div 3.2 = 13{,}200 \text{ ppmV}$$

Discussion. The actual concentration of the extracted vapor would be lower than 13,200 ppmV due to the fact that not all the air flows through the contaminated zone and that limitations of mass transfer were not considered in the above calculations.

V.1.3 Radius of influence and pressure profile

Selecting the number and locations of vapor extraction wells is one of the major tasks in design of in situ soil vapor extraction systems. The decisions are typically based on the radius of influence (R_I), which can be defined as the distance from the extraction well where the pressure drawdown is very small (P @ $R_I \sim 1$ atm). The most accurate and site-specific R_I values should be determined from steady-state pilot testing. The pressure drawdown data at the extraction well and the observation wells can be plotted as a function of the radial distance from the extraction well on a semilog plot to determine the R_I of that well. The approach is similar to the distance-drawdown method for aquifer tests, as described in Section II.3.3. The R_I is commonly chosen to be the distance where the pressure drawdown is less than 1% of the vacuum in the extraction well.

The field test data can also be analyzed by using the flow equations, which describe the subsurface air flow. The subsurface is usually heterogeneous, and the air flow through it can be very complex. As a simplified approximation, a flow equation was derived for a fully confined radial gas flow system in a permeable formation having uniform and constant properties.[3-6] References 3 through 6 are the basis for most of the sections on soil venting.

For the steady-state radial flow subject to the boundary conditions ($P = P_w$ @ $r = R_w$ and $P = P_{atm}$ @ $r = R_I$), the pressure distribution in the subsurface can be derived as

$$P_r^2 - P_w^2 = (P_{RI}^2 - P_w^2)\frac{\ln(r/R_w)}{\ln(R_1/R_w)} \qquad \text{[Eq. V.1.2]}$$

P_r = pressure at a radial distance r from the vapor extraction well
P_w = pressure at the vapor extraction well
P_{RI} = pressure at the radius of influence (= atmospheric pressure or a preset value)
r = radial distance from the vapor extraction well
R_I = radius of influence where pressure is equal to a preset value
R_w = well radius of the vapor extraction well

Eq. V.1.2 can be used to determine the R_I of a vapor extraction well if the pressure drawdown data of the extraction well and a monitoring well (or data of two monitoring wells) are known. As shown, the flow rate and the permeability of the formation are not included in this equation. The R_I

can also be estimated from the vapor extraction rate and the pressure drawdown data in the extraction well (see Section V.1.4).

If no pilot tests are conducted, an estimate is often made based on previous experiences. The R_I values ranging from 30 ft (9 m) to 100 ft (30 m) are reported in the literature, and typical pressures in the extraction wells range from 0.90 to 0.95 atm.[5] Shallower wells, less permeable subsurface, and lower applied vacuum in the extraction well generally correspond to smaller R_I values.

Example V.1.3A *Determine the radius of influence of a soil venting well by using the pressure drawdown data (pressure data are given in the atmospheric unit)*

Determine the radius of influence of a soil venting well with the following information:

Pressure at the extraction well = 0.9 atm
Pressure at a monitoring well 30 ft away from the venting well = 0.98 atm
Diameter of the venting well = 4 in

Solution:

a. Let us define the R_I as the location where P is equal to the atmospheric pressure. The R_I can be found by using Eq. V.1.2 as

$$P_r^2 - P_w^2 = (P_{RI}^2 - P_w^2)\frac{\ln(r / R_w)}{\ln(R_I / R_w)}$$

$$(0.98)^2 - (0.9)^2 = (1.0^2 - 0.9^2)\frac{\ln[30/(2/12)]}{\ln[R_I/(2/12)]}$$

$$R_I = 118 \text{ ft}$$

b. For comparison, let us now define the R_I as the location where the drawdown is equal to 1% of the vacuum in the extraction well. The vacuum in the extraction well = 1 – 0.9 = 0.1 atm. Thus, P_{RI} = 1 – (0.1)(1%) = 0.999 atm.

$$(0.98)^2 - (0.9)^2 = (0.999^2 - 0.9^2)\frac{\ln[30/(2/12)]}{\ln[R_I/(2/12)]}$$

$$R_I = 110 \text{ ft}$$

Discussion. The R_I value from (b), 110 ft, is about 7% shorter than that from (a), and it is a more realistic value.

Example V.1.3B *Determine the radius of influence of a soil venting well by using the pressure drawdown data (pressure data are given in inches of water)*

Determine the radius of influence of a soil venting well using the following information:

Pressure at the extraction well = 48 in water vacuum
Pressure at a monitoring well 40 ft away from the extraction well = 8 in water vacuum
Diameter of the vapor extraction well = 4 in

Strategy. The pressure data are expressed in inches of water. We need to convert them to the atmospheric unit or convert the atmospheric pressure to inches of water. A pressure of one atmosphere is equivalent to 33.9 feet of water column.

Solution:

a. Pressure at the extraction well = 48″ water vacuum = 33.9 – (48/12) = 29.9 ft of water = (29.9/33.9) = 0.88 atm.
Pressure at the monitoring well = 8″ water (vacuum) = 33.9 – (8/12) = 33.23 ft of water = (33.23/33.9) = 0.98 atm.
b. Let us define the R_I as the location where P is equal to the atmospheric pressure. The R_I can be found by using Eq. V.1.2 as

$$(0.98)^2 - (0.88)^2 = (1.0^2 - 0.88^2)\frac{\ln[40/(2/12)]}{\ln[R_I/(2/12)]}$$

$$R_I = 128 \text{ ft}$$

Example V.1.3C *Estimate the pressure drawdown in a soil venting monitoring well*

Using the pressure drawdown data given in Example V.1.3B, estimate the pressure drawdown (vacuum) in a monitoring well which is 20 ft away from the extraction well.

Strategy. Example V.1.3B gives the pressure drawdown data at (1) the monitoring well (P = 0.88 atm), (2) 40 ft away from the monitoring well (P

= 0.98 atm), and (3) the R_I (P = 1 atm). We can use any two of these three to estimate the pressure drawdown in a well that is 20 ft away from the extraction well.

Solution:

a. First, use the data of the extraction well and the monitoring well (r = 40 ft). The pressure at the monitoring well (r = 20 ft) can be found by using Eq. V.1.2 as

$$P_r^2 - (0.88)^2 = (0.98^2 - 0.88^2)\frac{\ln[20/(2/12)]}{\ln[40/(2/12)]}$$

$$P_r = 0.968 \text{ atm} = 10.0 \text{ in. of water (vacuum)}$$

b. We can also use the data of the extraction well and the R_I. The pressure at the monitoring well (r = 20 ft) can be found by using Eq. V.1.2 as

$$P_r^2 - (0.88)^2 = (1.0^2 - 0.88^2)\frac{\ln[20/(2/12)]}{\ln[128/(2/12)]}$$

$$P_r = 0.968 \text{ atm} = 10.0 \text{ in. of water (vacuum)}$$

c. We can also use the data of the monitoring well (r = 40 ft) and the R_I. The pressure at the monitoring well (r = 20 ft) can be found by using Eq. V.1.2 as

$$P_r^2 - (0.98)^2 = (1.0^2 - 0.98^2)\frac{\ln[20/40]}{\ln[128/40]}$$

$$P_r = 0.968 \text{ atm} = 10.0 \text{ in. of water (vacuum)}$$

Discussion. All three approaches yield the same result.

V.1.4 Vapor flow rates

The radial Darcian velocity, u_r, in homogeneous soil systems can be expressed as[4]

$$u_r = \left(\frac{k}{2\mu}\right)\frac{\left[\frac{P_w}{r\ln(R_w/R_I)}\right]\left[1-\left(\frac{P_{RI}}{P_w}\right)^2\right]}{\left\langle 1+\left[1-\left(\frac{P_{RI}}{P_w}\right)^2\right]\frac{\ln(r/R_w)}{\ln(R_w/R_I)}\right\rangle^{0.5}} \qquad \text{[Eq. V.1.3]}$$

where u_r is the vapor flow velocity at a radial distance "r" away from the well. The velocity at the wellbore, u_w, can be found by replacing r with R_w in the above equation as

$$u_w = \left(\frac{k}{2\mu}\right)\left[\frac{P_w}{R_w \ln(R_w / R_I)}\right]\left[1-\left(\frac{P_{RI}}{P_w}\right)^2\right] \qquad \text{[Eq. V.1.4]}$$

The volumetric vapor flow rate entering the extraction well, Q_w, can then be found as

$$Q_w = 2\pi R_w u_w H$$

$$= H\left(\frac{\pi k}{\mu}\right)\left[\frac{P_w}{\ln(R_w / R_I)}\right]\left[1-\left(\frac{P_{RI}}{P_w}\right)^2\right] \qquad \text{[Eq. V.1.5]}$$

where H is the perforation interval of the extraction well.

To convert the vapor flow rate entering the well to equivalent standard flow rates, Q_{atm} (where $P = P_{atm}$ = 1 atm), the following relationship can be used

$$Q_{atm} = \left(\frac{P_{well}}{P_{atm}}\right) Q_{well} \qquad \text{[Eq. V.1.6]}$$

Example V.1.4A *Estimate the extracted vapor flow rate of a soil venting well*

A soil venting well was installed at a site. Determine the radius of influence of this soil venting well using the following information:

Pressure at the extraction well = 0.9 atm
Pressure at a monitoring well 30 ft away from the venting well = 0.95 atm
Diameter of the venting well = 4 in

Calculate the steady-state flow rate entering the well per unit well screen length, vapor flow rate in the well, and the vapor rate at the extraction pump discharge by using the following additional information:

Permeability of the formation = 1 Darcy
Well screen length = 20 ft
Viscosity of air = 0.018 centipoise
Temperature of the formation = 20°C

Strategy. We need to perform a few unit conversions first:

1 atm = 1.013×10^5 N/m^2
1 Darcy = 10^{-8} cm^2 = 10^{-12} m^2
1 poise = 100 centipoise = 0.1 N/s/m^2

So, 0.018 centipoise = 1.8×10^{-4} poise = 1.8×10^{-5} N/s/m^2

Solution:

a. The radius of influence of the venting well has been determined in Example V.1.3A as 118 ft.
b. The radial air flow velocity at 20 ft away from the extraction well can be found by using Eq. V.1.3:

$$u_r = \left(\frac{k}{2\mu}\right)\frac{\left[\frac{P_w}{r\ln(R_w/R_I)}\right]\left[1-\left(\frac{P_{RI}}{P_w}\right)^2\right]}{\left\langle 1+\left[1-\left(\frac{P_{RI}}{P_w}\right)^2\right]\frac{\ln(r/R_w)}{\ln(R_w/R_I)}\right\rangle^{0.5}}$$

$$= \left[\frac{10^{-12}}{2(1.8\times10^{-5})}\right]\frac{\left[\frac{(0.9)(1.013\times10^5)}{(20)(0.3048)\ln[(2/12)/118]}\right]\left[1-\left(\frac{1}{0.9}\right)^2\right]}{\left\langle 1+\left[1-\left(\frac{1}{0.9}\right)^2\right]\frac{\ln[20/(2/12)]}{\ln[(2/12)/206]}\right\rangle^{0.5}}$$

$$= (2.78\times10^{-8})(-2.28\times10^3)(-0.2346)\div(1.17)^{0.5}$$

$$= 1.37\times10^{-5}\ \text{m/s} = 1.2\ \text{m/d}$$

c. The velocity at the wellbore, u_w, can be found by using Eq. V.1.4:

$$u_w = \left(\frac{k}{2\mu}\right)\left[\frac{P_w}{R_w\ln(R_w/R_I)}\right]\left[1-\left(\frac{P_{RI}}{P_w}\right)^2\right]$$

$$= \left[\frac{10^{-12}}{2(1.8\times10^{-5})}\right]\left[\frac{(0.9)(1.013\times10^5)}{(2/12)(0.3048)\ln[(2/12)/206]}\right]\left[1-\left(\frac{1}{0.9}\right)^2\right]$$

$$= (2.78\times10^{-8})(-2.73\times10^5)(-0.2346)$$

$$= 1.78\times10^{-3}\ \text{m/s} = 0.11\ \text{m/min} = 154\ \text{m/d}$$

d. The vapor flow rate entering the well per unit screen interval can be found by using Eq. V.1.5:

$$\frac{Q_w}{H} = 2\pi R_w u_w$$

$$= 2\pi[(2/12)(0.3048)\text{m}](0.11\text{m}/\text{min}) = 0.035\,\text{m}^3/\text{min}/\text{m}$$

e. The vapor flow rate in the well = (0.035 m^3/min/m)[(20 ft)(0.3048 m/ft)] = 0.21 m^3/min = 7.4 ft^3/min.

f. The vapor flow rate at the exhaust of the extraction pump can be calculated from Eq. V.1.6:

$$Q_{atm} = \left(\frac{P_{well}}{P_{atm}}\right)Q_{well} = \left(\frac{0.9}{1}\right)(0.21)$$

$$= 0.19\,\text{m}^3/\text{min} = 6.7\,\text{ft}^3/\text{min}$$

Discussion. Using consistent units in Eqs. V.1.3 and V.1.5 is very important. In the above calculations, the pressure is expressed in N/m^2, the distance in m, the permeability in m^2, and the viscosity in N/s/m^2. Consequently, the calculated velocity is in m/s.

Example V.1.4B *Estimate the radius of influence of a soil venting well by using the extracted vapor flow rate*

Determine the radius of influence of a soil venting well, with the following information:

Pressure at the venting well = 0.7 atm
Flow rate measured at the extraction pump discharge = 0.21 m^3/min
Well screen length = 5 m
Diameter of the venting well = 0.1 m
Permeability of the formation = 0.5 Darcy
Viscosity of air = 1.8×10^{-4} poise
Temperature of the formation = 20°C

Strategy. This problem can be viewed as the reverse of Example V.1.4A in which the radius of influence was given for estimation of the vapor extraction flow rate. In this problem, the flow rate was given to estimate the radius of influence. As in the previous example, a few unit conversions need to be performed first:

1 atm = 1.013×10^5 N/m^2

1 Darcy = 10^{-8} cm^2 = 10^{-12} m^2
1 poise = 100 centipoise = 0.1 N/s/m^2

So, 0.018 centipoise = 1.8×10^{-4} poise = 1.8×10^{-5} N/s/m^2.

Solution:

a. The vapor flow rate entering the well can be found by using Eq. V.1.6:

$$Q_{atm} = \left(\frac{P_{well}}{P_{atm}}\right) Q_{well} = 0.21 = \left(\frac{0.7}{1}\right) Q_{well}$$

$$Q_{well} = 0.30 \text{ m}^3/\text{min} = 0.005 \text{ m}^3/\text{s}$$

b. The radius of influence can be found by using Eq. II.1.5:

$$\frac{Q_w}{H} = \frac{0.005}{5}$$

$$= \left(\frac{\pi k}{\mu}\right)\left[\frac{P_w}{\ln(R_w / R_I)}\right]\left[1 - \left(\frac{P_{RI}}{P_w}\right)^2\right]$$

$$= \left(\frac{\pi(0.5 \times 10^{-12})}{1.8 \times 10^{-5}}\right)\left[\frac{(0.7)(1.013 \times 10^5)}{\ln(0.05 / R_I)}\right]\left[1 - \left(\frac{1}{0.7}\right)^2\right]$$

$$R_I = 31.9 \text{ m} \approx 32 \text{ m}$$

Discussion. Using consistent units is critical for successful calculations in this problem. Specifically, the flow rate is given in m^3/min, but it needs to be converted to m^3/s to match the viscosity units in Eq. V.1.5.

V.1.5 Contaminant removal rate

The contaminant removal rate ($R_{removal}$) can be determined by multiplying the extracted vapor flow rate (Q) with the vapor concentration (G):

$$R_{removal} = (G)(Q) \qquad \text{[Eq. V.1.7]}$$

Care should be taken to have G and Q in consistent units, and G should be in mass concentration units. Eq. V.1.1 can be used to estimate the initial vapor concentration if the free-product phase is present, while the procedure as illustrated in Example V.1.2C can be used to estimate the extracted vapor concentration in the absence of the free-product phase. It is worthwhile to note again that the calculated vapor concentrations from these procedures are the ideal and equilibrium values. The actual values should only be fractions of these values, mainly due to the facts that the entire air stream does not pass through the contaminated zone and that limitations of mass transfer exist (the system will not reach equilibrium in most, if not all, cases). Nevertheless, the calculated values provide useful information. One can compare them with the actual data from sampling and establish the correlation between them. The calculated data can then be calibrated, adjusted, and used for later predictions.

For example, if we know that only a fraction η of the air flows through the contaminated zone, Eq. V.1.7, should be modified as

$$R_{removal} = [(\eta)(G)](Q) \qquad \text{[Eq. V.1.8]}$$

The removal rate estimated from Eq. V.1.8 still represents the upper limit of the vapor concentration because it does not consider mass transfer limitations. The factor η can be considered as an overall efficiency factor if it takes into account the percentage of flow through the contaminated zone and the limitations of mass transfer.

The following procedure can be used to determine the contaminant removal rate:

Step 1: Determine the extraction vapor flow rate from field measurements or from the procedure described in Section V.1.4.
Step 2: Estimate the extracted vapor concentration using Eq. V.1.1 if the free-product phase is present, while the procedure illustrated in Example V.1.2C can be used to estimate the extracted vapor concentration in the absence of the free-product phase.
Step 3: Convert the vapor concentration into a mass concentration by using Eq. II.1.1.
Step 4: Adjust the calculated concentration from Step 2 by an overall efficiency factor, η.
Step 5: Calculate the mass removal rate by multiplying the flow rate (from Step 1) and the adjusted concentration (from Step 4).

Information needed for this calculation

- Extracted vapor flow rate, Q
- Extracted vapor concentration, G
- Overall efficiency factor relative to theoretical removal rate, η

Example V.1.5A Estimate the contaminant removal rate (in the presence of free-product phase)

Recently, a gasoline spill occurred at a gasoline station and caused subsurface contamination. A soil venting well was installed at the site for remediation.

The following data were obtained from the remedial investigation and a pilot test:

Pressure at the extraction well = 0.9 atm
Pressure at a monitoring well 30 ft away from the venting well = 0.95 atm
Diameter of the venting well = 4 in
Permeability of the formation = 1 Darcy
Well screen length = 20 ft
Viscosity of air = 0.018 centipoise
Temperature of the formation = 20°C

Estimate the contaminant removal rate at the beginning of the project.

Solution:

a. The pressure drawdown data are the same as those in Example V.1.4A, and the flow rate has been determined as 0.19 m^3/min, or 6.7 ft^3/min.
b. Assuming the free product phase is present, the saturated vapor concentration corresponding to the fresh gasoline is 340,000 ppmV, or 1343 g/m^3 (see Example V.1.2A). On the other hand, the saturated vapor concentration corresponding to the weathered gasoline is 49,000 ppmV, or 226 g/m^3.
c. Assuming the overall efficiency factor is equal to unity, the removal rate can be found from Eq. V.1.8 as

$$R_{removal} = [(\eta)(G)](Q) = [(1.0)(1343 \text{ g/m}^3)](0.19 \text{ m}^3\text{/min})$$
$$= 255 \text{ g/min} = 0.56 \text{ lb/min} = 809 \text{ lb/d (for the fresh gasoline)}$$
$$= [(1.0)(226 \text{ g/m}^3)](0.19 \text{ m}^3\text{/min}) = 42.9 \text{ g/min}$$
$$= 0.095 \text{ lb/min} = 136 \text{ lb/d (for the weathered gasoline)}$$

Discussion. The extracted vapor flow rate in this example is relatively small, at 6.7 ft^3/min. However, the calculated theoretical removal rates, 809 lb/d for the fresh gasoline and 136 lb/d for the weathered gasoline, are extraordinarily high. If the removal rate can be sustained at this level, the site would be cleaned up in a matter of days. Unfortunately, this is not the case. It normally takes months, if not longer, for a typical soil venting project to reach completion. This example illustrates that the theoretical equilibrium vapor concentration is higher than the practical values. For the case of gasoline, which is a mixture of compounds, the removal rate will drop as the more volatile ones leave the formation first (as indicated by the five times lower removal rate of the weathered gasoline). However, the value of 136 lb/d corresponding to the weathered gasoline is still on the high side because the limitations of mass transfer were not included in this calculation. The removal rate should drop further after the free-product phase disappears.

Example V.1.5B *Estimate the contaminant removal rate (in the absence of the free-product phase)*

A subsurface is contaminated with benzene. The average benzene concentration of the soil samples, taken from the contaminated zone, was 500 mg/kg. The following data were obtained from the remedial investigation and a pilot test:

Pressure at the extraction well = 0.9 atm
Pressure at a monitoring well 30 ft away from the venting well = 0.95 atm
Diameter of the venting well = 4 in
Permeability of the formation = 1 Darcy
Well screen length = 20 ft
Viscosity of air = 0.018 centipoise
Porosity = 0.35
Organic content = 0.03
Water saturation = 45%
Subsurface temperature = 25°C
Bulk density of soil = 1.7 g/cm^3

Estimate the contaminant removal rate at the beginning of the project.

Solution:

a. The pressure drawdown data are the same as those in Example V.1.4A, and the flow rate has been determined as 0.19 m^3/min, or 6.7 ft^3/min.
b. The subsurface data are the same as those in Example V.1.2D, and the extracted vapor concentration has been determined 42.3 mg/L, or 42.3 g/m^3.
c. Assuming the overall efficiency factor is equal to one, the removal rate can be found from Eq. V.1.8 as

$$R_{removal} = [(\eta)(G)](Q) = [(1.0)(42.3 \text{ g/m}^3)](0.19 \text{ m}^3/\text{min})$$
$$= 8.04 \text{ g/min} = 11{,}600 \text{ g/d} = 25.5 \text{ lb/d}$$

Discussion. The estimated value of 25.5 lb/d is on the high side because the overall efficiency factor is assumed to be unity. In addition, the removal rate would drop because the contaminant concentration in the subsurface decreases as the venting project progresses.

V.1.6 Cleanup time

Once the contaminant removal rate is determined, the cleanup time ($T_{cleanup}$) can be estimated as

$$T_{cleanup} = M_{spill} / R_{removal} \qquad \text{[Eq. V.1.9]}$$

where M_{spill} is the amount of spill to be removed. M_{spill} can be found by using the equation below:

$$M_{spill} = (X_{initial} - X_{cleanup})(M_S) = (X_{initial} - X_{cleanup})[(V_S)(\rho_b)] \quad \text{[Eq. V.1.10]}$$

where $X_{initial}$ is the average initial contaminant concentration in soil, $X_{cleanup}$ is the soil cleanup level, M_s is the mass of the contaminated soil, V_s is the volume of the contaminated soil, and ρ_b is the bulk density of the soil. If the cleanup level is very low compared to the initial contaminant concentration, it can be deleted from Eq. V.1.10 as a factor of safety for design.

The above equations appear simple. However, the estimation of cleanup time is complicated by the fact that the contaminant removal rate is changing. The rate decreases as the amount of the contaminants left in the soil decreases. One approach is to divide the cleanup into several time intervals. The removal rate for each interval is determined and used to estimate the cleanup time for each interval. The total cleanup can then be derived from summing the cleanup time of each interval. The following steps detail this approach:

Step 1: Determine the maximum possible contaminant concentration in soil in the absence of free-product, $X_{free\text{-}product}$ (see Example V.1.2C). If the average concentration of the soil samples exceeds this value, the free-product phase is present. Go to Step 2. If the average concentration of the samples is smaller, the free product phase is absent. Go to Step 5.

Step 2: Estimate the extracted vapor concentration using Eq. V.1.1 and then calculate the mass removal rate using Eq. V.1.8.

Step 3: Determine the amount of contaminants to be removed before the disappearance of the free product phase by using modified Eq. V.1.10 as

$$\begin{aligned} M_{removal} &= (X_{initial} - X_{free-product})(M_S) \\ &= (X_{initial} - X_{free-product})[(V_S)(\rho_b)] \end{aligned} \quad \text{[Eq. V.1.11]}$$

Step 4: Determine the required time for removal of the free product by using data from Steps 2 and 3 and Eq. V.1.9.

Step 5: Divide the ($X_{free\ product} - X_{cleanup}$) value into a few intervals. Use the average X of each interval to estimate the vapor concentration (see Example V.1.2D) and then calculate the mass removal rate using V.1.8. If no free-product phase is present initially, replace $X_{free\ product}$ with $X_{initial}$ in this step.

Step 6: Determine the amount of contaminants to be removed in each interval by using modified Eq. V.1.10:

$$M_{removal} = (X_{initial} - X_{final})(M_S) = (X_{initial} - X_{final})[(V_S)(\rho_b)] \quad \text{[Eq. V.1.12]}$$

The $X_{initial}$ and X_{final} here are the beginning and the end concentrations of each interval, respectively.

Step 7: Determine the required cleanup time for each interval by using data from Steps 5 and 6 and Eq. V.1.9.

Step 8: Sum the required time for each interval to calculate the total cleanup time.

Information needed for this calculation

- Contaminant concentrations of soil samples
- Henry's constant of the compound
- Organic water partition coefficient, K_{ow}
- Organic content, f_{oc}
- Porosity, ϕ
- Degree of water saturation
- Bulk density of soil, ρ_b

Example V.1.6A *Estimate the cleanup time (in the presence of free-product phase)*

Recently, a gasoline spill occurred at a gasoline station and caused subsurface contamination. A soil venting well was installed at the site for remediation. The following data were obtained from the remedial investigation and a pilot test:

Pressure at the extraction well = 0.9 atm
Pressure at a monitoring well 30 ft away from the venting well = 0.95 atm
Diameter of the venting well = 4 in
Permeability of the formation = 1 Darcy
Well screen length = 20 ft
Viscosity of air = 0.018 centipoise
Temperature of the formation = 20°C
Porosity = 0.35
Organic content in soil = 0.01
Degree of water saturation = 40%
Bulk density of soil = 1.8 g/cm^3
The size of the contaminant plume = 6500 ft^3
Initial average contaminant concentration in soil = 6000 mg/L
Required cleanup level = 100 mg/L
Overall efficiency factor relative to theoretical removal rate = 0.11

Estimate the required cleanup time.

Solution:

a. The pressure drawdown data are the same as those in Example V.1.4A, and the flow rate has been determined as 0.19 m^3/min, or 6.7 ft^3/min.

Presence of free-product phase

b. Determine the maximum possible contaminant concentration in the soil in the absence of free-product, $X_{free\text{-}product}$ (use the procedure illustrated in Example V.1.2C). Since no Henry's constant and K_{ow} data are available for gasoline, we use those of toluene, one of the common gasoline components, as an approximation. Use Table II.3.B to convert the Henry's constant to a dimensionless value:

$$H^* = H/RT = (6.7)/[(0.082)(273 + 20)] = 0.28 \text{ (dimensionless)}$$

Use Eq. II.3.14 to find K_{oc}:

$$K_{oc} = 0.63K_{ow} = 0.63\ (10^{2.73}) = (0.63)(537) = 338$$

Use Eq. II.3.12 to find K_p:

$$K_p = f_{oc}K_{oc} = (0.01)\ (338) = 3.4 \text{ L/kg}$$

Use the saturated gasoline vapor concentration of weathered gasoline, 226 mg/L (from Example V.1.2A) and Eq. II.3.23 to estimate $X_{free\text{-}product}$:

$$\frac{M_t}{V} = \left[\frac{(\phi_w)}{H} + \frac{(\rho_b)K_p}{H} + (\phi_a)\right]G$$

$$= \left[\frac{(0.35)(40\%)}{0.28} + \frac{(1.8)(3.4)}{0.28} + (0.35)(1-40\%)\right](226) = 5100 \text{ mg/L}$$

Divide this value by the bulk density of the soil to express the soil concentration in mg/kg:

$$X_{free\text{-}product} = 5100 \text{ mg/L} \div 1.8 \text{ kg/L} = 2830 \text{ mg/kg}$$

c. Determine the amount of contaminants to be removed before the disappearance of the free product phase by using Eq. V.1.11:

$$M_s = (V_s)(\rho_b) = (6500 \text{ ft}^3)[(1.8 \times 62.4 \text{ lb/ft}^3)]$$
$$= 730{,}100 \text{ lb} = 332{,}000 \text{ kg}$$

$$M_{removal} = (X_{initial} - X_{free\text{-}product})(M_s) = (6000 - 2830 \text{ mg/kg})(332{,}000 \text{ kg})$$
$$= 9.40 \times 10^8 \text{ mg} = 940 \text{ kg}$$

d. Estimate the extracted vapor concentration using Eq. V.1.1. As determined in Example V.1.2A, the saturated gasoline vapor concentrations are 1343 mg/L and 226 mg/L for the fresh and the weathered gasoline, respectively. Since the observed VOC concentrations of the extracted vapor often decrease exponentially over time, the geometric average of these two values is used as the average concentration for this interval:

$$G = \sqrt{(1343)(226)} = 551 \text{ mg/L}$$

e. Calculate the mass removal rate using Eq. V.1.8:

$$R_{removal} = [(\eta)(G)](Q) = [(0.11)(551 \text{ g/m}^3)](0.19 \text{ m}^3/\text{min})$$
$$= 11.5 \text{ g/min} = 16.6 \text{ kg/d}$$

f. Determine the required cleanup time by using data from (c) and (e) and Eq. V.1.9:

$$T_1 = M_{removal} \div R_{removal} = (940 \text{ kg}) \div 16.6 \text{ kg/d} = 56.6 \text{ days}$$

Absence of free-product phase

g. At the end of the free-product removal, the contaminant concentration in soil is 2830 mg/kg, corresponding to a theoretical vapor concentration of 226 mg/L. The cleanup level of soil for this project is 100 mg/kg. The average of 2830 and 100 is equal to 1465. To estimate the required cleanup time, we divide it into two intervals. The first interval is the time required to reduce the concentration from 2830 to 1465 mg/kg and the other is from 1465 to 100 mg/kg.

h. Determine the amount of contaminants to be removed in the first interval by using Eq. V.1.12 as

$$M_{removal} = (X_{initial} - X_{final})(M_s) = (2830 - 1465 \text{ mg/kg})(332{,}000 \text{ kg})$$
$$= 4.53 \times 10^8 \text{ mg} = 453 \text{ kg}$$

For this interval the initial theoretical vapor concentration is 226 mg/L (corresponding to 2830 mg/kg), the final theoretical vapor concentration (corresponding to 1465 mg/kg) can be easily found as

$$G_{final} = 226 \times (1465/2830) = 117 \text{ mg/L}$$

The geometric average of these two is used as the average concentration for this interval:

$$G = \sqrt{(226)(117)} = 163 \text{ mg/L}$$

Calculate the mass removal rate by using Eq. V.1.8:

$$R_{removal} = [(\eta)(G)](Q) = [(0.11)(163 \text{ g/m}^3)](0.19 \text{ m}^3\text{/min})$$
$$= 3.4 \text{ g/min} = 4.9 \text{ kg/d}$$

Determine the required cleanup time by using Eq. V.1.9:

$$T_2 = M_{removal} \div R_{removal} = (453 \text{ kg}) \div 4.9 \text{ kg/d} = 92.4 \text{ days}$$

i. In the second interval the amount of the contaminant mass to be removed is the same as that of the first interval, 453 kg. The initial theoretical vapor concentration is 117 mg/L (corresponding to 1465 mg/kg), the final theoretical vapor concentration (corresponding to 100 mg/kg) can be easily found as

$$G_{final} = 117 \times (100/1465) = 8 \text{ mg/L}$$

The geometric average of these two is used as the average concentration for this interval:

$$G = \sqrt{(117)(8)} = 30.6 \text{ mg/L}$$

Calculate the mass removal rate by using Eq. V.1.8:

$$R_{removal} = [(\eta)(G)](Q) = [(0.11)(30.6 \text{ g/m}^3)](0.19 \text{ m}^3\text{/min})$$
$$= 0.64 \text{ g/min} = 0.92 \text{ kg/d}$$

Determine the required cleanup time by using Eq. V.1.9:

$$T_3 = M_{removal} \div R_{removal} = (453 \text{ kg}) \div 0.92 \text{ kg/d} = 492 \text{ days}$$

Entire project. The total cleanup time required

$$= \text{T1} + \text{T2} + \text{T3} = 56.6 + 92.4 + 492 = 641 \text{ days.}$$

Discussion

1. For the three intervals, the average mass removal rates drop significantly from 16.6 kg/d in the first interval to 0.92 kg/d in the third interval.
2. For the two intervals during the absence of free product, the second interval takes 492 days and the first interval takes only 92 days to remove the same amount of contaminant.
3. The cleanup time of 641 days is not acceptable in most project applications. One may consider increasing the extraction flow rate or adding more wells.

4. Only two intervals were used to analyze the period between free-product disappearance and final cleanup; the estimate would be more accurate if more intervals were used.
5. If the free-product phase is not present initially, solve the problem by starting from part (g).

V.1.7 Effect of temperature on soil venting

In a soil venting project, the subsurface temperature will affect both the air flow rate and the vapor concentration. At a higher temperature, the vapor pressure of an organic compound would be higher. On the other hand, the higher subsurface temperature will yield a lower air flow rate because air viscosity increases with temperature:

$$\frac{\mu @ T_1}{\mu @ T_2} = \sqrt{\frac{T_1}{T_2}} \qquad \text{[Eq. V.1.13]}$$

where T is the subsurface temperature, expressed in Kelvin or Rankine units. From Eq. V.1.5, the ratio of the flow rates at two temperatures will be

$$\frac{Q @ T_1}{Q @ T_2} = \sqrt{\frac{T_2}{T_1}} \qquad \text{[Eq. V.1.14]}$$

As shown in the above equation, the vapor flow rate will be lower at higher temperatures. However, since the vapor concentration will be much higher at higher temperature, the mass removal rate will still be higher at higher temperatures.

Example V.1.7 *Estimate the extracted vapor flow rate of a soil venting well at elevated temperatures*

A soil venting well was installed at a site. The following data were obtained from the remedial investigation:

Pressure at the extraction well = 0.9 atm
Pressure at a monitoring well 30 ft away from the venting well = 0.95 atm
Diameter of the venting well = 4 in
Permeability of the formation = 1 Darcy
Well screen length = 20 ft
Viscosity of air = 0.018 centipoise
Temperature of the formation = 20°C

The extracted vapor flow rate has been estimated, as shown in Example V.1.4, to be 6.7 ft^3/min under the above conditions. If the subsurface temperature is raised to 30°C, what will be the vapor flow rate (if all the other conditions are kept the same)?

Solution:
The new air flow rate can be found by using Eq. V.1.12 as

$$\frac{Q@30°C}{Q@20°C} = \sqrt{\frac{273.2+20}{273.2+30}}$$

$$Q\ @\ 30°C = (6.7)(0.967) = 6.5\ ft^3/min$$

Discussion. The temperature affects the air flow rate insignificantly. For a 10°C increase in temperature, the flow rate decreases by less than 4%.

V.1.8 Number of vapor extraction wells

There are three main considerations in determining the number of vapor extraction wells necessary for a soil venting project. First, a successful soil venting project should have sufficient extraction wells to cover the entire area of contamination. In other words, the entire contaminated zone should be within the influence of the wells, thus

$$N_{wells} = \frac{1.2(A_{contamination})}{\pi R_I^2} \quad \text{[Eq. V.1.15]}$$

The factor of 1.2 is arbitrarily chosen to account for the overlapping of the influence areas among the wells as well as the fact that the peripheral wells may reach outside the contaminated zone. Second, the number of wells should be sufficient to complete the site cleanup within an acceptable time frame.

$$R_{acceptable} = \frac{M_{spill}}{T_{cleanup,acceptable}} \quad \text{[Eq. V.1.16]}$$

$$N_{wells} = \frac{R_{acceptable}}{R_{removal}} \quad \text{[Eq. V.1.17]}$$

The minimum number of the vapor extraction wells should be the larger of the two that are determined from Eqs. V.1.15 and V.1.17.

The last, and probably the most important, consideration is the economical one. There is a trade-off between the number of wells and the total treatment cost.

Example V.1.8 *Determine the number of venting wells required*

For the soil venting project described in Example V.1.6A, it is desired to clean up the site in 9 months. Determine the number of venting wells needed. The plume has a cross-sectional area of 850 ft^2.

Solution:

a. The flow rate from one venting well has been determined as 0.19 m^3/min, or 6.7 ft^3/min. At this flow rate, the cleanup will take 641 days. To meet the 9-month cleanup schedule, the removal rate should be 2.4 (= 641 ÷ 270) times faster. Therefore, we need to increase the flow rate by 2.4 times or to have 3 venting wells.
b. The radius of influence of one venting well has been determined from Example V.1.3A as 118 ft. The number of wells needed to cover the plume can be determined by using Eq. V.1.15 as

$$N_{wells} = \frac{1.2(A_{contamination})}{\pi R_I^2} = \frac{1.2(850)}{\pi(118)^2} = 0.23$$

Therefore, one well should be enough to cover the entire plume, unless the plume has a very long stripe shape.
c. Based on the above results, three venting wells would be required.

V.1.9 *Sizing of vacuum pump (blower)*

The theoretical horse-power requirements ($hp_{theoretical}$) of vacuum pumps, blowers, or compressors for an ideal gas undergoing an isothermal compression (PV = constant) can be expressed as[7]

$$hp_{theoretical} = 3.03 \times 10^{-5} P_1 Q_1 \ln \frac{P_2}{P_1} \qquad \text{[Eq.V.1.18]}$$

where P_1 = intake pressure, lb_f/ft^2, P_2 = final delivery pressure, lb_f/ft^2, and Q_1 = air flow rate at the intake condition, ft^3/min.

For an ideal gas undergoing an isentropic compression (PV^k = constant), the following equation is applicable for single-stage compressors:[7]

$$hp_{theoretical} = \frac{3.03 \times 10^{-5} k}{k-1} P_1 Q_1 \left[\left(\frac{P_2}{P_1} \right)^{(k-1)/k} - 1 \right] \qquad \text{[Eq.V.1.19]}$$

where k is the ratio of specific heat of gas at constant pressure to specific heat of gas at constant volume. For the typical soil venting applications, it is appropriate to use k = 1.4.

For reciprocating compressors, the efficiencies (E) are generally in the range of 70 to 90% for isentropic and 50 to 70% for isothermal compression. The actual horsepower requirement can be found as

$$hp_{actual} = \frac{hp_{theoretical}}{E} \qquad \text{[Eq.V.1.20]}$$

Example V.1.9 *Determine the required horsepower of the vacuum pump in soil venting*

Two vapor extraction wells are installed. The design flow rate of each well is 40 ft³/min, and the design well head pressure is 0.9 atm. A vacuum pump is to serve both wells. Estimate the required horsepower of the vacuum pump.

Solution:

a. The pressure in the well, P_1 = 0.9 atm = (0.9)(14.7) = 13.2 psi = (13.2)(144) psf = 1905 lb/ft².
b. Assuming isothermal expansion, use Eq. V.1.18 to determine the theoretical power requirement as

$$hp_{theoretical} = 3.03 \times 10^{-5} P_1 Q_1 \ln \frac{P_2}{P_1} = 3.03 \times 10^{-5} P_2 Q_2 \ln \frac{P_2}{P_1}$$

$$= 3.03 \times 10^{-5} [(14.7)(144)][(2)(40)] \ln \frac{(14.7)(144)}{(1905)} = 0.54hp$$

Assuming an isothermal efficiency of 60%, the actual horsepower required is determined by using Eq. VI.3.8 as

$$hp_{actual} = \frac{hp_{theoretical}}{E} = \frac{0.54}{60\%} = 0.9hp$$

c. Assuming isentropic expansion, use Eq. V.1.19 to determine the theoretical power requirement as

$$hp_{theoretical} = \frac{3.03 \times 10^{-5} k}{k-1} P_1 Q_1 \left[\left(\frac{P_2}{P_1} \right)^{(k-1)/k} - 1 \right]$$

$$= \frac{(3.03 \times 10^{-5})(1.4)}{1.4 - 1} [(14.7)(144)][(40)(2)] \left[\left(\frac{(14.7)(144)}{1905} \right)^{(1.4-1)/1.4} - 1 \right] = 0.55hp$$

Assuming an isentropic efficiency of 80%, the actual horsepower required is determined by using Eq. V.1.20 as

$$hp_{actual} = \frac{hp_{theoretical}}{E} = \frac{0.55}{80\%} = 0.7hp$$

Discussion. In general, the energy necessary for an isentropic compression is greater than that for an equivalent isothermal compression. In soil venting applications, the difference between the inlet and final discharge pressures is relatively small. Consequently, the theoretical power requirements for isothermal and isentropic compression are very similar, as illustrated in this example.

V.2 Soil bioremediation

V.2.1 Description of the soil bioremediation process

Soil bioremediation utilizes microorganisms or their metabolic products to degrade organic contaminants in soil. Soil bioremediation can be conducted under aerobic or anaerobic conditions, but aerobic bioremediation is more popular. The final products of complete aerobic biodegradation are carbon dioxide and water.

Bioremediation may also be either in situ or ex situ. Ex situ soil bioremediation processes are more developed and demonstrated than in situ processes. Ex situ bioremediation is typically performed in one of three common systems: (1) static soil pile, (2) in vessel, and (3) slurry bioreactor. The static soil pile is the most popular format. The method uses excavated soil stockpiled on the treatment site with perforated pipes embedded in the pile as the conduit for air supply. To improve process and emission control, the soil piles are usually covered.

In situ treatment enhances the natural microbial activity of undisturbed soil in place to decompose organic contaminants. A nutrient solution is often percolated or injected into the subsurface to support the growth of the organics–degraders. Run-on and run-off controls for moisture control and waste containment are often required. In a slurry bioreactor, soil is mixed with a nutrient solution under controlled operating conditions.

Microorganisms require moisture, oxygen (or absence of oxygen), nutrients, and a suitable set of environmental factors to grow. The environmental factors include pH, temperature, and absence of toxic conditions. Table V.2.A summarizes the critical conditions for bioremediation.

V.2.2 Moisture requirement

As shown in Table V.2.A, the optimal moisture content for soil bioremediation is 25 to 85% of the water-holding capacity. In most cases, soil moisture

Table V.2.A Critical Conditions for Bioremediation

Environmental factor	Optimum conditions
Available soil water	25–85% water holding capacity
Oxygen	*Aerobic metabolism:* > 0.2 mg/L dissolved oxygen, air-filled pore space to be > 10% by volume
	Anaerobic metabolism: oxygen concentration to be < 1% by volume
Redox potential	*Aerobes and facultative anaerobes:* >50 millivolts
	Anaerobes: < 50 millivolts
Nutrients	Sufficient N, P, and other nutrients (suggested C:N:P molar ratio of 120:10:1)
pH	5.5–8.5 (for most bacteria)
Temperature	15–45°C (for mesophiles)

From U.S. EPA, Site Characterization for Subsurface Remediation, EPA/625/R-91/026, Office of Research and Development, U.S. EPA, Washington, DC, 1991.

will be below or in the lower end of this range; therefore, addition of moisture is commonly needed.

The moisture present in the vadose zone is often quantified by a term called the volumetric water content or degree of saturation. Volumetric water content varies from zero to the value of porosity, while degree of saturation varies from zero to one and refers to the percentage of pore space occupied by moisture. For complete saturation, the volumetric water content is equal to porosity, and the degree of water saturation is 100%. The following formula can be used to determine the volume of water needed for bioremediation.

$$\text{Volume of water needed, } V_{water} = (\text{volume of soil})$$

$$(\text{desired moisture content} - \text{initial moisture content}) \qquad \text{[Eq. V.2.1]}$$

$$= (V_{soil})(\phi_{w,f} - \phi_{w,i}) = (V_{soil})[(\phi)(S_{w,f} - S_{w,i})]$$

where $\phi_{w,i}$ = initial soil moisture content, $\phi_{w,f}$ = desired soil moisture content, ϕ = porosity of soil, $S_{w,i}$ = initial degree of saturation, and $S_{w,f}$ = desired degree of saturation.

Example V.2.2 *Determine the moisture requirement for soil bioremediation*

A UST-removal project resulted in a 375-yd^3 gasoline-contaminated soil pile that has to be treated before disposal. Bioremediation has been selected as the treatment method. Determine the amount of water needed for the first spray.

Use the following simplified assumptions in your calculation:

1. Porosity of soil = 35%
2. Initial degree of saturation = 20%

Solution:

a. Based on Table V.2.A, the optimal moisture content for soil bioremediation is 25 to 85% of the water-holding capacity. Without conducting an optimization study, the middle value of this range, 60%, is selected.
b. Water needed = (375)[(0.35)(60% − 20%)] = 52.5 yd^3 = 1417.5 ft^3 = 10,600 gal.

Discussion. Addition of make-up water is often needed periodically. The frequency of moisture additions depends heavily on the climate of the project site.

V.2.3 Nutrient requirements

Nutrients for microbial activity usually exist in the subsurface. However, with the presence of organic contaminants, additional nutrients are often needed to support the bioremediation. The nutrients to enhance microbial growth are assessed primarily on the nitrogen and phosphorus requirements of the microorganisms. As shown in Table V.2.A, the suggested C:N:P ratio is 120:10:1. (Some other references suggest C:N:P = 100:10:1.) The ratio is on a molar basis. It means that every 120 moles of carbon requires 10 moles of nitrogen and 1 mole of phosphorous. For bioremediation, a feasibility study is always recommended. Determination of an optimal nutrient ratio should be part of the feasibility study. If no other information is available, the ratio mentioned above can be used. The example in this section will show that the amount of nutrients needed is relatively small, and so is the cost. Nutrients are often dissolved in water first and then applied to the soil by spraying or irrigation.

To determine the nutrient requirements, the following procedure can be used:

Step 1: Determine the mass of the organics present in the contaminated soil.
Step 2: Divide the mass of organics by its molecular weight to find the moles of the contaminant.
Step 3: Multiply the moles of contaminant from step 2 by the number of C in the compound's formula.
Step 4: Determine the moles of nitrogen and phosphorus needed using the optimal C:N:P ratio. For example, if the ratio is C:N:P = 120:10:1, then moles of nitrogen needed = (moles of carbon present) × (10/120) and moles of phosphorus needed = (moles of carbon present) × (1/120).
Step 5: Determine the amount of nutrient needed.

Information needed for this calculation

- The mass of the organic contaminants
- The chemical formula of the contaminants
- The optimal C:N:P ratio
- The chemical formula of the nutrients

Example V.2.3 *Determine the nutrient requirement for soil bioremediation*

The results of a feasibility study indicate that the excavated soil in a stockpile (Example V.2.2) is suitable for on-site above-ground bioremediation. The feasibility study also determined the optimum C:N:P molar ratio to be 100:10:1. Estimate the amount and cost of nutrients (in lbs) needed to bioremediate the gasoline contamination.

Use the following assumptions in your calculation:

a. Volume of excavated soil in pile = 375 yd^3
b. Initial mass of gasoline in the pile = 158 kg
c. Soil porosity = 0.35
d. Formula of gasoline (assumed) = C_7H_{16}
e. The amounts of N and P naturally occurring in the excavated soil is insignificant
f. Trisodium phosphate ($Na_3PO_4 \cdot 12H_2O$) as the P source; price = \$3/lb
g. Ammonium sulfate ($(NH_4)_2SO_4$) as the N source; price = \$1/lb
h. One-time nutrient addition only

Solution:

a. Determine the number of moles of gasoline. Molecular weight of gasoline = 7 × 12 + 1 × 16 = 100 and moles of gasoline = 158/100 = 1.58 kg-mole.
b. Determine the number of moles of C in soil. Since there are seven carbon atoms in each gasoline molecule, as indicated by its formula, C_7H_{16}, then

$$\text{Moles of C} = (1.58)(7) = 11.06 \text{ kg-mole}$$

c. Determine the number of moles of N needed (using the C:N:P ratio).
Mole of N needed = (10/100)(11.06) = 1.106 kg-mole
Mole of $(NH_4)_2SO_4$ needed = 1.106/2 = 0.553 kg-mole (each mole of ammonium sulfate contains two moles of N).
Amount of $(NH_4)_2SO_4$ needed = (0.553)[(14 + 4)(2) + 32 + (16)(4)] = 73 kg = 161 lbs = \$161.
d. Determine the number of moles of P needed (using the C:N:P ratio).
Mole of P needed = (1/100)(11.06) = 0.111 kg-mole

Mole of $Na_3PO_4 \cdot 12H_2O$ needed = 0.111 kg-mole.
Amount of $Na_3PO_4 \cdot 12H_2O$ needed = (0.111)[(23)(3) + 31 + (16)(4) + (12)(18)] = 42 kg = 92.5 lbs = $277.

Discussion. The cost of nutrients is relatively low compared to other parts of the project expenses.

V.2.4 *Oxygen requirement*

For soil bioremediation, the oxygen involved in the biological activity is often supplied through the oxygen in the air. There are plenty of oxygen compounds in the air. Oxygen is approximately 21% by volume in our ambient air. On the other hand, oxygen is sparingly soluble in water. At 20°C, the saturated dissolved oxygen (DO) in the water is only about 9 mg/L.

Let us use the following simplified scheme to demonstrate the oxygen requirements:

	C	+	O_2	→	CO_2
Moles	1		1		1
Mass (gram or lb)	12		32		44

The above equation illustrates that each mole of carbon element requires one mole of oxygen molecule, or every 12 g of carbon requires 32 g of oxygen, a ratio of 2.67. Other elements in the contaminants, such as hydrogen, nitrogen, and sulfur, would also demand oxygen for bioremediation. For example, the theoretical amount of oxygen required to aerobically biodegrade benzene can be found as

	C_6H_6	+	$7.5O_2$	→	$6CO_2$	+	$3H_2O$
Moles	1		7.5		6		3
Mass (gram or lb)	78		240		264		54

This indicates that each mole of benzene requires 7.5 moles of oxygen molecule, or every 78 g of carbon requires 240 g oxygen, a ratio of 3.08, which is larger than 2.67 based on pure carbon. Using benzene as the basis, it means that every gram of hydrocarbon requires approximately 3 g of oxygen for aerobic degradation. It should be noted that this is the theoretical ratio based on the stoichiometric relationship. A higher oxygen concentration would enhance the rate of biodegradation. Using this ratio, the amount of oxygen in an aqueous solution saturated with oxygen can only support biodegradation of contaminants at a concentration of 3 mg/L or less. (The saturated dissolved oxygen concentration is 9 mg/L at 20°C, and the DO concentration in a typical aquifer would be much lower than this value.)

Example V.2.4A Determine the oxygen concentration in air

Determine the mass concentration of oxygen in ambient air at 20°C. Express the answer in the following units: mg/L, g/L, and lb/ft³.

Solution:

The oxygen concentration in the ambient air is approximately 21% by volume, which is equal to 210,000 ppmV. Eq. II.1.1 or II.1.2 can be used to convert it to a mass concentration:

$$1\,\text{ppmV} = \frac{\text{MW}}{24.05}[\text{mg/m}^3] \quad \text{at } 20°\text{C}$$

$$= \frac{32}{24.05} = 1.33\,\text{mg/m}^3 = 0.00133\,\text{mg/L} \qquad \text{[Eq. II.1.1]}$$

or

$$1\,\text{ppmV} = \frac{MW}{385} \times 10^{-6}\ [\text{lb/ft}^3] \quad \text{at } 68°\text{F}$$

$$= \frac{32}{385} \times 10^{-6} = 0.083 \times 10^{-6}\,\text{lb/ft}^3 \qquad \text{[Eq. II.1.2]}$$

Therefore,

$$210{,}000\ \text{ppmV} = (210{,}000)(0.00133\ \text{mg/L}) = 279\ \text{mg/L} = 0.28\ \text{g/L}$$
$$= (210{,}000)(0.083 \times 10^{-6}) = 0.0175\ \text{lb/ft}^3.$$

Discussion. The oxygen concentration in the ambient air, 279 mg/L, is much higher than the saturated dissolved oxygen (DO) concentration in water, 9 mg/L at 20°C.

Example V.2.4B Determine the necessity of oxygen addition for in situ soil bioremediation

A subsurface is contaminated with 5000 mg/L of gasoline. The air in the subsurface is relatively stagnant. The bulk density of the soil is 1.8 g/cm³, the degree of water saturation in the soil is 30%, and the porosity is 40%.

Demonstrate that the oxygen in the soil void is not sufficient to support the complete biodegradation of the intruding gasoline contaminants.

Solution:

Basis = 1 m³ of soil.

a. Determine the mass of the contaminants present. Mass of the soil matrix = (1 m^3)(1800 kg/m^3) = 1800 kg
Mass of the contaminants

$$= (5000 \text{ mg/kg})(1800 \text{ kg}) = 9{,}000{,}000 \text{ mg} = 9000 \text{ g}$$

b. Use the 3.08 ratio to determine the oxygen requirements for complete oxidation. Oxygen requirement = (3.08)(9000) = 27,720 g.
c. Determine the amount of oxygen in the soil moisture (assuming that the moisture is saturated with oxygen and the saturated dissolved oxygen concentration in water at 20°C is approximately 9 mg/L).
The volume of the soil moisture = $V\phi S_w$ = (1 m^3)(40%)(30%) = 0.12 m^3 = 120 L.
The amount of oxygen in soil moisture = $(V_l)(DO)$ = (120 L)(9 mg/L) = 1080 mg = 1.08 g.
d. Determine the amount of oxygen in air (assuming that the oxygen concentration in the pore void is the same as that in the ambient air, 21% by volume, or 279 mg/L from example V.2.4A).
The volume of the air void, $V_{air\ void} = V\phi(1 - S_w)$ = (1 m^3)(40%)(1 – 30%) = 0.28 m^3 = 280 L.
The amount of oxygen in air void = $(V_{air\ void})(G_{oxygen})$ = (280 L)(279 mg/L) = 78,120 mg = 78.1 g.
e. The total available oxygen in the soil moisture and the air void = 1.08 + 78.1 = 79.2 g/m^3 soil << 27,720 g/m^3 soil.

Discussion
1. The amount of available oxygen in the soil moisture, 1.08 g/m^3 soil, is much smaller than that in the air void, 78.1 g/m^3.
2. It would need at least 255 (= 27,720/78.1) void volumes of fresh air to supply sufficient oxygen for complete biodegradation. The minimum fresh air requirement = (255)($V_{air\ void}$) = (255)(280 L/m^3 soil) = 71,400 L/m^3 soil = 71.4 m^3 fresh air/m^3 soil.

V.3 Soil washing/solvent extraction/soil flushing

V.3.1 Description of the soil washing process

The majority of the organic and inorganic contaminants attached to soil are adsorded onto small clay or silt particles that have large specific surface areas. These small clay and silt particles, in turn, attach to larger sand and gravel particles by compaction and adhesion. In this section, three remediation technologies are described: soil washing, solvent extraction, and soil flushing. These are similar in that they use solvents to extract or separate contaminants from the soil matrix.

Soil washing is a water-based washing process. The major removal mechanisms include (1) the desorption of contaminants from the soil, and consequent dissolution in the washing fluid and (2) suspension of the clay and silt particles with bound contaminants into the wash water. The contaminants are readily washed off from sand and gravel, which often account for a large portion of the soil matrix. Separating the sand and gravel from the heavily contaminated clay and silt particles greatly reduces the volume of contaminated soil. Soil washing makes further treatment or disposal much easier.

Various chemicals can be added to the aqueous solution to enhance the desorption or dissolution of the contaminants. For example, an acidic solution is often used to extract heavy metals from contaminated soils. Solvent extraction is identical to soil washing, except that organic solvents rather than aqueous solutions are employed to extract organic contaminants from soil. Commonly used solvents include alcohol and liquefied propane and butane. Supercritical fluids are also used.

Soil flushing differs from soil washing or solvent extraction in that it is an in situ process in which water or solvent flushes the contaminated zone to desorb or dissolve the contaminants. The elutriate is then collected from the wells or drains for further treatment.

A mass balance equation can be written to relate the contaminant concentrations in the soil before and after washing with the contaminant concentration in the spent washing fluid (assuming that the fresh washing fluid does not contain any contaminants) as

$$X_{initial}M_s = X_{final}M_s + CV_l \qquad \text{[Eq. V.3.1]}$$

where $X_{initial}$ = initial contaminant concentration in the soil (mg/kg), X_{final} = final contaminant concentration in the soil (mg/kg), M_s = mass of soil washed (kg), C = contaminant concentration in the spent washing fluid (mg/L), and V_l = volume of the washing fluid used (L).

The term on the left-hand side of Eq. V.3.1 represents the total contaminant mass before washing, and the terms on the right-hand side represent the mass left on the soil and the mass dissolved in the liquid phase at the end of washing, respectively. If an equilibrium condition is achieved at the end of the washing, the contaminant concentration in the soil and that in the liquid can be related by the partition equation described in Chapter two:

$$X_{final} = K_pC \qquad \text{[Eq. V.3.2]}$$

where K_p is the partition equilibrium constant. By inserting Eq. V.3.2 into Eq. V.3.1, the relationship between the initial and final contaminant concentrations of the soil can be expressed by Eq. V.3.3. or Eq. V.3.4 as

$$\frac{X_{final}}{X_{initial}} = \frac{1}{1+\left(\dfrac{V_l}{M_sK_p}\right)} \qquad \text{[Eq. V.3.3]}$$

$$X_{final} = \frac{1}{1+\left(\frac{V_l}{M_s K_p}\right)} \times X_{initial} \qquad \text{[Eq. V.3.4]}$$

For several washers in series, the final contaminant concentration can be determined by

$$\frac{X_{final}}{X_{initial}} = \frac{1}{1+\left(\frac{V_{1,1}}{M_s K_p}\right)} \times \frac{1}{1+\left(\frac{V_{1,2}}{M_s K_p}\right)} \times \frac{1}{1+\left(\frac{V_{1,3}}{M_s K_p}\right)} \times \ldots \qquad \text{[Eq. V.3.5]}$$

Example V.3.1A *Determine the efficiency of soil washing*

A sandy subsurface is contaminated with 500 mg/L of 1,2-dichloroethane (DCA) and 500 mg/L of pyrene. Soil washing is proposed to remediate the soil. A batch washer that can accommodate 1000 kg of soil is designed. For each batch of operation, 1000 gal of clean water is used as the washing fluid. Determine the final concentrations of the two contaminants in the soil.

Use the following data from the site assessment in design:

Bulk density of soil = 1.8 g/cm^3
Fraction of organic carbon of aquifer materials = 0.005
$K_{oc} = 0.63K_{ow}$

Solution:

a. From Table II.3.C

$$\text{Log}(K_{ow}) = 1.53 \text{ for 1,2-DCA} \rightarrow K_{ow} = 34$$

$$\text{Log}(K_{ow}) = 4.88 \text{ for pyrene} \rightarrow K_{ow} = 75{,}900$$

b. Using the given relationship, $K_{oc} = 0.63K_{ow}$, we obtain

$$K_{oc} = (0.63)(34) = 22 \text{ (for 1,2-DCA)}$$

$$K_{oc} = (0.63)(75{,}900) = 47{,}800 \text{ (for pyrene)}$$

c. Using Eq. II.3.12, $K_p = f_{oc}K_{oc}$, and $f_{oc} = 0.005$, we obtain

$$K_p = (0.005)(22) = 0.11 \text{ L/kg (for 1,2-DCA)}$$

$$K_p = (0.005)(47{,}800) = 239 \text{ L/kg (for pyrene)}$$

d. Use Eq. V.3.5 to find the final concentration as (1000 gal = 3785 L):

$$X_{\text{final}} = \frac{1}{1+\left(\dfrac{V_l}{M_s K_p}\right)} \times X_{\text{initial}}$$

$$X_{final} = \frac{1}{1+\left(\dfrac{3{,}785}{(1000)(0.11)}\right)} \times 500 = 14.1 \text{ mg/L} \quad \text{for 1,2-DCA}$$

$$X_{final} = \frac{1}{1+\left(\dfrac{3{,}785}{(1000)(239)}\right)} \times 500 = 492 \text{ mg/L} \quad \text{for pyrene}$$

Discussion

1. Pyrene is very hydrophobic and its K_p value is very high. This example demonstrates that water washing is essentially ineffective in removing pyrene from soil. Other washing fluids such as organic solvents should be considered instead.
2. The calculated values are based on an assumption that the liquid and the soil are in equilibrium. For a practical reactor design, an equilibrium condition is seldom reached. Consequently, the actual final concentration would be higher.

Example V.3.1B *Determine the efficiency of soil washing (two washers in series)*

The single washer described in Example V.3.1A could not reduce the 1,2-DCA concentration to below 10 mg/L. An engineer proposed using two smaller washers in series. The washer still accommodates 1000 kg of soil, but only 500 gallons of fresh water is added to each washer. Can this system meet the cleanup requirements?

Solution:

Use Eq. V.3.6 to find the final concentration for two washers in series as ($V_{l,1}$ = $V_{l,2}$ = 500 gal = 1893 L):

$$X_{final} = \frac{1}{1+\left(\dfrac{V_{l,1}}{M_s K_p}\right)} \times \frac{1}{1+\left(\dfrac{V_{l,2}}{M_s K_p}\right)} \times X_{final}$$

$$= \frac{1}{1+\left(\dfrac{1893}{(1000)(0.11)}\right)} \times \frac{1}{1+\left(\dfrac{1893}{(1000)(0.11)}\right)} \times 500 = 1.5 \text{mg/L}$$

Discussion

1. In both cases, same amount of water, 1000 gal is used for 1000 kg of soil. However, use of two washers in series yields a lower final concentration.
2. The calculated values are based on an assumption that the liquid and the soil are in equilibrium. For a practical reactor design, an equilibrium condition is seldom reached. Consequently, the actual final concentration would be higher.

V.4 Low-temperature heating (desorption)

V.4.1 Description of the low temperature heating (desorption) process

In the low-temperature heating (desorption) process, volatile and semivolatile contaminants are removed from soils, sediments, or slurries through volatilization that is enhanced by elevated temperatures. The process is typically operated at temperatures from 200 up to 1000°F. The term "low temperature" is used to differentiate the process from incineration. At these lower temperatures, the contaminants are physically driven off from the soil matrix instead of being combusted. The produced off-gas requires further treatment before being vented to the atmosphere.

V.4.2 Design of the low-temperature heating (desorption) process

There are no set guidelines for design of a low-temperature heating reactor. The time required to achieve a specific final concentration would depend mainly on the following factors:

1. Temperature inside the reactor: the higher the temperature, the higher the desorption rate will be and, consequently, the shorter the retention time.
2. Mixing conditions inside the reactor: better mixing conditions will enhance the heat transfer and improve venting of the desorbed contaminants.
3. Volatility of the contaminants: the more volatile the contaminants are, the shorter the required retention time will be.
4. Size of the soil: the smaller the soil particles, the easier the desorption will be.
5. Types of soil: clay has a stronger affinity with contaminants, and, thus, the contaminants will be more difficult to desorb from clayey material.

The rate of desorption or the required detention time to remediate a specific type of soil to a permissible concentration can be best determined from a pilot study. The results from the pilot study should then be used for preliminary design of full-scale operation. The desorption process can be conducted in a batch mode or in a continuous mode. For the continuous mode, the reactor can be modeled as a CFSTR, if the soil is relatively well-mixed inside the reactor. For the desorption reaction, a first-order type of reaction is a reasonable assumption. For a first-order reaction, the relationship among the influent and final concentrations, reaction rate constant, and retention time are as follows (See Chapter four for more detailed discussions):

Batch reactor

$$\frac{C_f}{C_i} = e^{-k\tau} \quad \text{or} \quad C_f = (C_i)e^{-k\tau} \qquad \text{[Eq. V.4.1]}$$

CFSTR

$$\frac{C_{out}}{C_{in}} = \frac{1}{1 + k(V/Q)} = \frac{1}{1 + k\tau} \qquad \text{[Eq. V.4.2]}$$

Example V.4.2A Determine the residence time for low-temperature heating (batch mode of operation)

A batch-type, low-temperature-heating soil reactor is proposed to treat soil contaminated with 2500 mg/kg of total petroleum hydrocarbon (TPH). A pilot study was conducted, and it took 25 minutes to reduce the concentration to 150 mg/kg. First-order kinetics apply. If the required final soil TPH concentration is 50 mg/kg, what should be the design residence time of the soil in the reactor?

Solution:

a. Determine the rate constant by using Eq. V.4.1:

$$\frac{C_f}{C_i} = e^{-k\tau} = \frac{150}{2500} = e^{-k(25)}$$

So,

$$k = 0.113/\text{min}$$

b. Now, we use this rate constant and Eq. V.4.1 to determine the required retention time:

$$\frac{C_f}{C_i} = e^{-k\tau} = \frac{50}{2500} = e^{-(0.113)\tau}$$

So,

$$\tau = 35 \text{ minutes}$$

Discussion. The rate constant is often obtained from bench-scale experiments by using the batch-type reactors.

Example V.4.2B *Determine the residence time for low-temperature heating (continuous mode of operation)*

A low-temperature heating soil reactor is proposed to treat soil contaminated with 3000 mg/kg of total petroleum hydrocarbon (TPH) in a continuous mode of operation. Assume that the reactor is a CFSTR and that first-order kinetics apply. A pilot study was conducted and the reaction rate constant was determined to be 0.3/min. The required final soil TPH concentration is 100 mg/kg.

a. What should be the design residence time of the soil in the reactor?
b. The soil content of the reactor is to be kept at less than 30% of the total reactor volume to allow for efficient mixing. Estimate the required size of the reactor vessel to treat the contaminated soil at a rate of 500 kg/hr.

Solution:

a. Determine the required retention time by using Eq. V.4.2:

$$\frac{C_{out}}{C_{in}} = \frac{1}{1+k\tau} = \frac{100}{3000} = \frac{1}{1+(0.3)\tau}$$

$1 + 0.3\tau = 30$. So, $\tau = 97$ min $= 1.61$ hr.

b. Assuming the bulk density of soil in the reactor is 1.5 g/cm^3, the volumetric feeding rate of the soil can be found as

$$Q_{soil} = (500 \text{ kg/hr}) \div 1.5 \text{ kg/L} = 333 \text{ L/hr}$$

The minimum reactor size can be found from the definition of the retention time as

$$\tau = V/Q = 1.61 \text{ hr} = V/(333 \text{ L/hr})$$

So,

$$V = 537 \text{ L}$$

With the soil occupying less than 30% of the total reactor volume, the required reactor volume ($V_{reactor}$) can be found as

$$V_{reactor} = (537) \div 30\% = 1790 \text{ L} = 473 \text{ gal}$$

References

1. U.S. EPA, Site Characterization for Subsurface Remediation, EPA/625/R-91/026, Office of Research and Development, U.S. EPA, Washington, DC, 1991.
2. U.S. EPA, Control of Air Emission from Superfund Sites, EPA/625/R-91/012, Office of Research and Development, U.S. EPA, Washington, DC, 1992.
3. Johnson, P. C. and Ettinger, R. A., Considerations for the design of in situ vapor extraction systems: radius of influence vs. zone of remediation, *Ground Water Monitor. Rev.*, Summer, 1994.
4. Johnson, P. C., Kemblowski, M. W., and Colthart, J. D., Qualitative analysis for the cleanup of hydrocarbon-contaminated soils by in-situ soil venting, *Groundwater*, 28(3), 413, 1992.
5. Johnson, P. C., Stanely, C. C., Kemblowski, M. W., Byers, D. L., and Colthart, J. D., A practical approach to the design, operation, and monitoring of in situ soil-venting systems, *Ground Water Monitor. Rev.*, Spring, 199b.
6. Kuo, J. F., Aieta, E. M., and Yang, P. H., Three-dimensional soil venting model and its applications, in *Emerging Technologies in Hazardous Waste Management II*, Tedder, D. W. and Pohland, F. G., Eds., American Chemical Society Symposium Series 468, American Chemical Society, Washington, DC, 1991, 382–400.
7. Peters, M. S. and Timmerhaus, K. D., *Plant Design and Economics for Chemical Engineers*, 4th ed., McGraw-Hill, New York, 1991.

chapter six

Groundwater remediation

This chapter starts with design calculations for capture zone and optimal well spacing. The rest of the chapter focuses on design calculations for commonly used in situ and ex situ groundwater remediation techniques, including bioremediation, air sparging, air stripping, advanced oxidation processes, and activated carbon adsorption.

VI.1 Hydraulic control (groundwater extraction)

When a groundwater aquifer is contaminated, groundwater extraction is often needed. Groundwater extraction through pumping mainly serves two purposes: (1) to minimize the plume migration or spreading and (2) to reduce the contaminant concentrations in the impacted aquifer. The extracted water often needs to be treated before being injected back into the aquifer or released to surface water bodies. *Pump and treat* is a general term used for groundwater remediation that removes contaminated groundwater and treats it above ground.

Groundwater extraction is typically accomplished through one or more pumping or extraction wells. Pumping of groundwater stresses the aquifer and creates a cone of depression or a capture zone. Choosing appropriate locations for the pumping wells and spacing among the wells is an important component in design. Pumping wells should be strategically located to accomplish rapid mass removal from areas of the groundwater plume where contaminants are heavily concentrated. On the other hand, they should be located to allow full capture of the plume to prevent further migration. In addition, if containment is the only objective for the groundwater pumping, the extraction rate should be established at a minimum rate sufficient to prevent the plume migration. (The more the groundwater is extracted, the higher the treatment cost.) On the other hand, if groundwater cleanup is required, the extraction rate may need to be enhanced to shorten the remediation time. For both cases, major questions to be answered for design of a groundwater pump-and-treat program are

1. What is the optimum number of pumping wells required?
2. Where would be the optimal locations of the extraction wells?
3. What would be the size (diameter) of the wells?
4. What would be the depth, interval, and size of the perforations?
5. What would be the construction materials of the wells?
6. What would be the optimum pumping rate for each well?
7. What would be the optimal treatment method for the extracted groundwater?
8. What would be the disposal method for the treated groundwater?

This section will illustrate common design calculations to determine the influence of a pumping well. The results from these calculations can provide answers to some of the above questions.

VI.1.1 Cone of depression

When a groundwater extraction well is pumped, the water level in its vicinity will decline to provide a gradient to drive water toward the well. The gradient is steeper as the well is approached, and this results in a cone of depression. In dealing with groundwater contamination problems, evaluation of the cone of depression of a pumping well is critical because it represents the limit that the well can reach.

The equations describing the steady-state flow of an aquifer from a fully penetrating well have been discussed earlier in Section III.2. The equations were used in that section to estimate the drawdown in the wells as well as the hydraulic conductivity of the aquifer. These equations can also be used to estimate the radius of influence of a groundwater extraction well or to estimate the groundwater pumping rate. This section will illustrate these applications.

Steady-state flow in a confined aquifer

The equation describing steady-state flow of a confined aquifer (an artesian aquifer) from a fully penetrating well is shown below. A fully penetrating well means that the groundwater can enter at any level from the top to the bottom of the aquifer.

$$Q = \frac{Kb(h_2 - h_1)}{528 \log(r_2 / r_1)} \quad \text{for American Practical Units}$$

$$= \frac{2.73\, Kb(h_2 - h_1)}{\log(r_2 / r_1)} \quad \text{for SI} \qquad \text{[Eq. VI.1.1]}$$

where Q = pumping rate or well yield (in gpm or m^3/d), h_1, h_2 = static head measured from the aquifer bottom (in ft or m), r_1, r_2 = radial distance from the pumping well (in ft or m), b = thickness of the aquifer (in ft or m), and K = hydraulic conductivity of the aquifer (in gpd/ft^2 or m/d).

Example VI.1.1A *Radius of influence from pumping a confined aquifer*

A confined aquifer 30 ft (9.1 m) thick has a piezometric surface 80 ft (24.4 m) above the bottom confining layer. Groundwater is being extracted from a 4-in (0.1 m) diameter fully penetrating well.

The pumping rate is 40 gpm (0.15 m^3/min). The aquifer is relatively sandy with a hydraulic conductivity of 200 gpd/ft^2. Steady-state drawdown of 5 ft (1.5 m) is observed in a monitoring well 10 ft (3.0 m) from the pumping well. Determine

a. The drawdown in the pumping well
b. The radius of influence of the pumping well

Solutions:

a. First let us determine h_1 (at r_1 = 10 ft):

$$h_1 = 80 - 5 = 75 \text{ ft} \quad (\text{or} = 24.4 - 1.5 = 22.9 \text{ m})$$

To determine the drawdown at the pumping well, set r at the well = well radius = (2/12) ft = 0.051 m and use Eq. VI.1.1:

$$40 = \frac{(200)(30)(h_2 - 75)}{528 \log[(2/12)/10]} \rightarrow h_2 = 68.7 \text{ ft}$$

or

$$[(0.15)(1440)] = \frac{2.73[(200)(0.0410)](9.1)(h_2 - 22.9)}{\log(0.051/3.0)} \rightarrow h_2 = 21.0 \text{ m}$$

So, the drawdown in the pumping well = 80 – 68.7 = 11.3 ft (or = 24.4 – 21.0 = 3.4 m).

b. To determine the radius of influence of the pumping well, set r at the radius of influence (r_{RI}) to be the location where the drawdown is equal to zero. We can use the drawdown information of the pumping well as

$$40 = \frac{(200)(30)(68.7 - 80)}{528 \log[(2/12)/r_{RI}]} \rightarrow r_{RI} = 270 \text{ ft}$$

or

$$[(0.15)(1440)] = \frac{2.73[(200)(0.0410)](9.1)(21.0 - 24.4)}{\log(0.051/r_{RI})} \rightarrow r_{RI} = 82 \text{ m}$$

Similar results can also be derived from using the drawdown information of the observation well as

$$40 = \frac{(200)(30)(75 - 80)}{528 \log[10/r_{RI}]} \rightarrow r_{RI} = 263 \text{ ft}$$

or

$$[(0.15)(1440)] = \frac{2.73[(200)(0.0410)](9.1)(22.9 - 24.4)}{\log(3/r_{RI})} \rightarrow h_2 = 78 \text{ m}$$

Discussion

1. In (a), 0.041 is the conversion factor to convert the hydraulic conductivity from gpd/ft^2 to m/day. The factor was taken from Table III.1.A.
2. Calculations in (a) have demonstrated that the results would be the same by using two different systems of units.
3. The "$h_1 - h_2$" term can be replaced by "$s_2 - s_1$," where s_1 and s_2 are the drawdown values at r_1 and r_2, respectively.
4. The differences in the calculated r_{RI} values in (b) come mainly from the unit conversions and data truncations.

Example VI.1.1B *Estimate the groundwater extraction rate of a confined aquifer from steady-state drawdown data*

Use the following information to estimate the groundwater extraction rate of a pumping well in a confined aquifer:

Aquifer thickness = 30.0 ft (9.1 m) thick
Well diameter = 4-in (0.1 m) diameter
Well perforation depth = full penetrating
Hydraulic conductivity of the aquifer = 400 gpd/ft^2
Steady-state drawdown = 2.0 ft observed in a monitoring well 5 ft from the pumping well = 1.2 ft observed in a monitoring well 20 ft from the pumping well

Solutions:

Inserting the data into Eq. VI.1.1, we obtain

$$Q = \frac{Kb(h_2 - h_1)}{528 \log(r_2/r_1)} = \frac{(400)(30)(2.0 - 1.2)}{528 \log(20/5)} = 30.2 \text{ gpm}$$

Discussion. The "$h_1 - h_2$" term can be replaced by "$s_2 - s_1$," where s_1 and s_2 are the drawdown values at r_1 and r_2, respectively.

Example VI.1.1C Estimate the pumping rate from a confined aquifer

Determine the rate of discharge (in gpm) of a confined aquifer being pumped by a fully penetrating well. The aquifer is composed of medium sand. It is 90 ft thick with a hydraulic conductivity of 550 gpd/ft^2. The drawdown of an observation well 50 ft away is 10 ft, and the drawdown in a second observation well 500 ft away is 1 ft.

Solution:

This problem is very similar to Ex. VI.1.1B. The flow rate can be calculated by using Eq. VI.1.1 as

$$Q = \frac{Kb(H-h)}{528\log(R/r)} = \frac{Kb(h_2-h_1)}{528\log(r_2/r_1)}$$

$$= \frac{(550)(90)[(90-1)-(90-10)]}{(528)\log(500/50)} = 844 \text{ gpm}$$

Steady-state flow in an unconfined aquifer

The equation describing the steady-state flow of an unconfined aquifer (water-table aquifer) from a fully penetrating well can be expressed as

$$Q = \frac{K(h_2^2-h_1^2)}{1055\log(r_2/r_1)} \quad \text{for American Practical Units}$$

$$= \frac{1.366\,K(h_2^2-h_1^2)}{\log(r_2/r_1)} \quad \text{for SI} \qquad \text{[Eq. VI.1.2]}$$

All the terms are as defined for Eq. VI.1.1.

Example VI.1.1D Radius of influence from pumping an unconfined aquifer

A water-table aquifer is 40 ft (12.2 m) thick. Groundwater is being extracted from a 4-inch (0.1 m) diameter fully penetrating well.

The pumping rate is 40 gpm (0.15 m^3/min). The aquifer is relatively sandy with a hydraulic conductivity of 200 gpd/ft^2. Steady-state drawdown of 5 ft (1.5 m) is observed in a monitoring well at 10 ft (3.0 m) from the pumping well. Estimate

a. The drawdown in the pumping well
b. The radius of influence of the pumping well

Solutions:

a. First let us determine h_1 (at $r_1 = 10$ ft):

$$h_1 = 40 - 5 = 35 \text{ ft} \quad (\text{or} = 12.2 - 1.5 = 10.7 \text{ m})$$

To determine the drawdown at the pumping well, set r at the well = well radius = (2/12) ft = 0.051 m, and use Eq. VI.1.2:

$$40 = \frac{(200)(h_2^2 - 35^2)}{1055 \log[(2/12)/10]} \rightarrow h_2 = 29.2 \text{ ft}$$

or

$$[(0.15)(1440)] = \frac{1.366[(200)(0.0410)](h_2^2 - 10.7^2)}{\log(0.051/3.0)} \rightarrow h_2 = 9.0 \text{ m}$$

So, the drawdown in the extraction well = 40 – 29.2 = 10.8 ft (or = 12.2 – 9.0 = 3.2 m).

b. To determine the radius of influence of the pumping well, set r at the radius of influence (r_{RI}) to be the location where the drawdown is equal to zero. We can use the drawdown information of the pumping well as

$$40 = \frac{(200)(29.2^2 - 40^2)}{1055 \log[(2/12)/r_{RI}]} \rightarrow r_{RI} = 580 \text{ ft}$$

or

$$[(0.15)(1440)] = \frac{1.366[(200)(0.0410)](9.0^2 - 12.2^2)}{\log(0.051/r_{RI})} \rightarrow r_{RI} = 168 \text{ m}$$

Similar results can also be derived from using the drawdown information of the observation well as

$$40 = \frac{(200)(35^2 - 40^2)}{1055 \log[10/r_{RI}]} \rightarrow r_{RI} = 598 \text{ ft}$$

or

$$[(0.15)(1440)] = \frac{1.366[(200)(0.0410)](10.7^2 - 12.2^2)}{\log(3/r_{RI})} \rightarrow r_{RI} = 181 \text{ m}$$

Discussion

1. In Eq. VI.1. for confined aquifers, the "$h_1 - h_2$" term can be replaced by "$s_2 - s_1$," where s_1 and s_2 are the drawdown values at r_1 and r_2, respectively. However, no analogy can be made here, that is, "$h_2^2 - h_1^2$" in Eq. VI.1.2 cannot be replaced by "$s_1^2 - s_2^2$."
2. The differences in the calculated r_{RI} values in (b) come mainly from the unit conversions and data truncations.

Example VI.1.1E Estimate the groundwater extraction rate of an unconfined aquifer from steady-state drawdown data

Use the following information to estimate the groundwater extraction rate of a pumping well in an unconfined aquifer:

Aquifer thickness = 30.0 ft (9.1 m) thick
Well diameter = 4-in (0.1 m) diameter
Well perforation depth = full penetrating
Hydraulic conductivity of the aquifer = 400 gpd/ft^2
Steady-state drawdown = 2.0 ft observed in a monitoring well 5 ft from the pumping well = 1.2 ft observed in a monitoring well 20 ft from the pumping well

Solutions:

a. First we need to determine h_1 and h_2:

$$h_1 = 30.0 - 2.0 = 28.0 \text{ ft}$$

$$h_2 = 30.0 - 1.2 = 28.8 \text{ ft}$$

b. Inserting the data into Eq. VI.1.2, we obtain

$$Q = \frac{K(h_2^2 - h_1^2)}{1055 \log(r_2/r_1)} = \frac{400(28.8^2 - 28.0^2)}{1055 \log(20/5)} = 28.6 \text{ gpm}$$

VI.1.2 Capture zone analysis

One key element in design of a groundwater extraction system is selection of proper locations for the pumping wells. If only one well is used, the well should be strategically located to create a capture zone that encloses the entire contaminant plume. If two or more wells are used, the general interest is to find the maximum distance between any two wells such that no contaminants can escape through the interval between the wells. Once such distances are determined, one can depict the capture zone of these wells from the rest of the aquifer.

To delineate the capture zone of a groundwater pumping system in an actual aquifer can be a very complicated task. To allow for a theoretical approach, let us consider a homogeneous and isotropic aquifer with a uniform thickness and assume the groundwater flow is uniform and steady. The theoretical treatment of this subject starts from one single well and expands to multiple wells. The discussions are mainly based on the work by Javandel and Tsang.[2]

One groundwater extraction well

For easier presentation, let the extraction well be located at the origin of an x-y coordinate system (Figure VI.1.A). The equation of the dividing streamlines that separate the capture zone of this well from the rest of the aquifer (sometimes referred to as the "envelope") is

$$y = \pm \frac{Q}{2Bu} - \frac{Q}{2\pi Bu} \tan^{-1} \frac{y}{x} \qquad \text{[Eq. VI.1.3]}$$

where B = aquifer thickness (ft or m), Q = groundwater extraction rate (ft^3/s or m^3/s), and u = regional groundwater velocity (ft/s or m/s) = Ki.

Figure VI.1.A illustrates the capture zone of a single pumping well. The larger the Q/Bu value is (i.e., larger groundwater extraction rate, slower groundwater velocity, or shallower aquifer thickness), the larger the capture zone. Three interesting sets of x and y values of the capture zone:

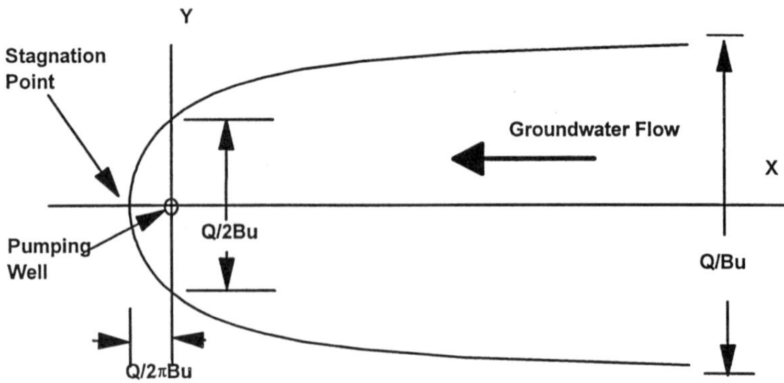

Figure VI.1.A Capture zone of a single well.

1. The stagnation point, where y is approaching zero,
2. The sidestream distance at the line of the extraction well, where x = 0, and
3. The asymptotic values of y, where $x = \infty$.

If these three sets of data are determined, the rough shape of the capture zone can be depicted. At the stagnation point (where y is approaching zero), the distance between the stagnation point and the pumping well is equal to $Q/2\pi Bu$, which represents the farthest downstream distance that the pumping well can reach. At $x = 0$, the maximum sidestream distance from the extraction well is equal to $\pm Q/4Bu$. In other words, the distance between the dividing streamlines at the line of the well is equal to $Q/2Bu$. The asymptotic value of y (where $x = \infty$) is equal to $\pm Q/2Bu$. Thus, the distance between the streamlines far upstream from the pumping well is Q/Bu.

Note that the parameter in Eq. VI.1.3 (Q/Bu) has a dimension of length. To draw the envelope of the capture zone, Eq. VI.1.3 can be rearranged as

$$x = \frac{y}{\tan\left\langle\left[+1-\left(\frac{2Bu}{Q}\right)y\right]\pi\right\rangle} \quad \text{for positive } y \text{ values} \quad \text{[Eq. VI.1.4A]}$$

$$x = \frac{y}{\tan\left\langle\left[-1-\left(\frac{2Bu}{Q}\right)y\right]\pi\right\rangle} \quad \text{for negative } y \text{ values} \quad \text{[Eq. VI.1.4B]}$$

A set of (x, y) values can be obtained from these equations by first specifying a value of y. The envelope is symmetrical about the x-axis.

Example VI.1.2A *Draw the envelope of a capture zone of a groundwater pumping well*

Delineate the capture zone of a groundwater recovery well with the following information:

Q = 60 gpm
Hydraulic conductivity = 2000 gpd/ft^2
Groundwater gradient = 0.01
Aquifer thickness = 50 ft

Solution:

a. Determine the groundwater velocity, u:

$$u = (K)(i) = [(2000 \text{ gal/d/ft}^2)(1 \text{ d/1440 min})(1 \text{ ft}^3/7.48 \text{ gal})](0.01)$$
$$= 1.86 \times 10^{-3} \text{ ft/min}$$

b. Determine the value of the parameter, *Q/Bu:*

$$\frac{Q}{Bu} = \frac{(60 \text{ gal/min})(1 \text{ ft}^3/7.48 \text{ gal})}{(50 \text{ ft})(1.8 \times 10^{-3} \text{ ft/min})}$$

or

$$= \frac{60 \text{ gal/min}}{(50 \text{ ft})[2000 \text{ gal/d/ft}^2)(1 \text{ d/1440 min})(0.01)]} = 86.4 \text{ ft}$$

c. Establish a set of the (x, y) values using Eq. VI.1.4. First specify values of y. Select smaller intervals for small y values. The following table lists some of the data points used to plot Figure E.VI.1.2A.

y (ft)	x (ft)
0	0.00
0.1	–13.74
1	–13.73
5	–13.14
10	–11.24
20	–2.34
30	21.01
40	168.78
–0.1	–13.74
–1	–13.73
–5	–13.14
–10	–11.24
–20	–2.34
–30	21.01
–40	168.78

Discussion

1. The capture zone curve is symmetrical about the x-axis as shown in the table or in the figure. Note that Eq. VI.1.4A should be used for positive y values and Eq. VI.1.4B for negative y values.
2. Do not specify the y values beyond the values of $\pm Q/2Bu$. As discussed, $\pm Q/2Bu$ are the asymptotic values of the capture zone curve ($x = \infty$).

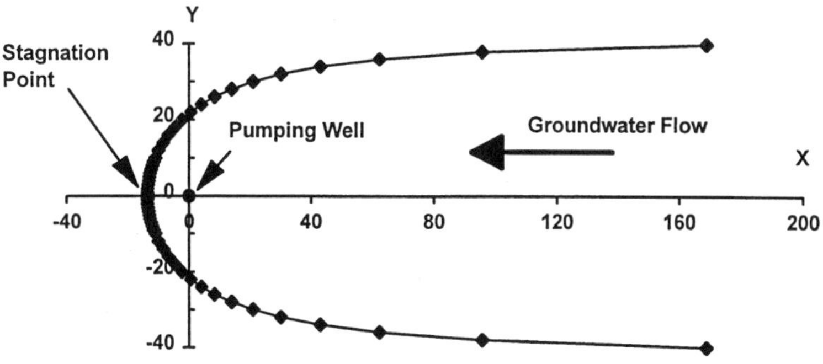

Figure E.VI.1.2A Capture zone of a single well.

Example VI.1.2B Determine the downstream and sidestream distances of a capture zone

A groundwater extraction well is installed in an aquifer (hydraulic conductivity = 1000 gpd/ft^2, gradient = 0.015, and aquifer thickness = 80 ft).

The design pumping rate is 50 gpm. Delineate the capture zone of this recovery well by specifying the following characteristic distances of the capture zone:

a. The sidestream distance from the well to the envelope of the capture zone at the line of the pumping well
b. The downstream distance from the well to the stagnation point of the envelope
c. The sidestream distance of the envelope far upstream of the pumping well

Solution:

a. Determine the groundwater velocity, u:

$$u = (K)(i) = [(1000\ \text{gal/d/ft}^2)(1\ \text{d}/1440\ \text{min})(1\ \text{ft}^3/7.48\ \text{gal})](0.015)$$
$$= 1.39 \times 10^{-3}\ \text{ft/min}$$

b. Determine the sidestream distance from the well to the envelope of the capture zone at the line of the pumping well, $Q/4Bu$:

$$\frac{Q}{4\,Bu} = \frac{(50\,\text{gal/min})(1\,\text{ft}^3/7.48\,\text{gal})}{(4)(80\,\text{ft})(1.39\times10^{-3}\ \text{ft/min})} = 15.0\,\text{ft}$$

c. Determine the downstream distance from the well to the stagnation point of the envelope, $Q/2\pi Bu$:

$$\frac{Q}{2\pi Bu} = \frac{(50\,\text{gal/min})(1\,\text{ft}^3/7.48\,\text{gal})}{(2)(\pi)(80\,\text{ft})(1.39\times10^{-3}\,\text{ft/min})} = 9.6\ \text{ft}$$

d. Determine the sidestream distance of the envelope far upstream of the pumping well, $Q/2Bu$:

$$\frac{Q}{2Bu} = \frac{(50\,\text{gal/min})(1\,\text{ft}^3/7.48\,\text{gal})}{(2)(80\,\text{ft})(1.39\times10^{-3}\,\text{ft/min})} = 30.0\,\text{ft}$$

e. The general shape of the envelope can be defined by using the above characteristic distances:

x (ft)	y (ft)	Note
0	0	Well location
−9.6	0	Downstream distance (stagnation point)
0	15	Sidestream distance at the line of the well
0	−15	Sidestream distance at the line of the well
150*	30	Sidestream distance at far upstream of the well
150*	−30	Sidestream distance at far upstream of the well

* The sidestream distance far upstream of the well, ±30 ft, should occur at $x = \infty$. A value of 150, which is ten times the sidestream distance at the line of well, is used here as the value of x.

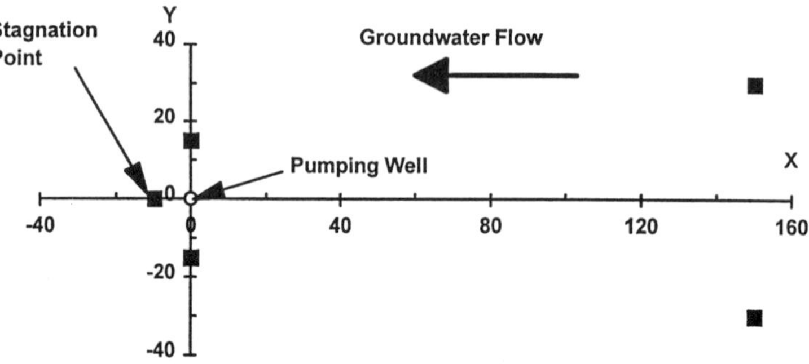

Figure E.VI.1.2B Capture zone of a single well.

Multiple wells

Table V1.1.A summarizes some characteristic distances of the capture zone for multiple groundwater monitoring wells located on a line perpendicular to the flow direction. As shown in the table, the distance between the dividing streamlines far upstream from the pumping wells is equal to $n(Q/Bu)$, where n is the number of the pumping wells. This distance is twice the distance between the streamlines at the line of the wells.

The downstream distance for multiple wells is very similar to that of the single pumping well, i.e., $Q/2\pi Bu$.

Table VI.1.A Characteristic Distances of the Capture Zone for Groundwater Pumping Wells

No. of extraction wells	Optimal distance between each pair of extraction wells	Distance between the streamlines at the line of the wells	Distance between the streamlines at far upstream from the wells
1	—	0.5 Q/Bu	Q/Bu
2	0.32 Q/Bu	Q/Bu	2 Q/Bu
3	0.40 Q/Bu	1.5 Q/Bu	3 Q/Bu
4	0.38 Q/Bu	2 Q/Bu	4 Q/Bu

Modified from Javandel, I. and Tsang, C.-F., *Groundwater*, 24(5), 616–625, 1986. With permission.

Example VI.1.2C Determine the downstream and sidestream distances of a capture zone for multiple wells

Two groundwater extraction wells are to be installed in an aquifer (hydraulic conductivity = 1000 gpd/ft^2, gradient = 0.015, and aquifer thickness = 80 ft).

The design pumping rate for each well is 50 gpm. Determine the optimal distance between the two wells and delineate the capture zone of these recovery wells by specifying the following characteristic distances of the capture zone:

a. The sidestream distance from the wells to the envelope of the capture zone at the line of the pumping wells
b. The downstream distance from the wells to stagnation points of the envelope
c. The sidestream distance of the envelope far upstream of the pumping wells

Solution:

a. Determine the groundwater velocity, u:

$$u = (K)(i) = [(1000\text{ gal/d/ft}^2)(1\text{ d}/1440\text{ min})(1\text{ ft}^3/7.48\text{ gal})](0.015)$$
$$= 1.39 \times 10^{-3}\text{ ft/min}$$

b. Determine the optimum distance between these two wells, 0.32 Q/Bu:

$$\frac{0.32Q}{Bu} = \frac{(0.32)(50\,\text{gal/min})(1\,\text{ft}^3/7.48\,\text{gal})}{(80\,\text{ft})(1.39\times10^{-3}\,\text{ft/min})} = 19.2\,\text{ft}$$

The distance of each well to the origin is half of this value = 0.16 Q/Bu = 9.6 ft.

c. Determine the sidestream distance from the well to the envelope of the capture zone at the line of the pumping well, $Q/2Bu$:

$$\frac{Q}{2\,Bu} = \frac{(50\,\text{gal/min})(1\,\text{ft}^3/7.48\,\text{gal})}{(2)(80\,\text{ft})(1.39 \times 10^{-3}\,\text{ft/min})} = 30.0\,\text{ft}$$

d. Determine the downstream distance from the well to the stagnation point of the envelope, $Q/2\pi Bu$:

$$\frac{Q}{2\,\pi Bu} = \frac{(50\,\text{gal/min})(1\,\text{ft}^3/7.48\,\text{gal})}{(2)(\pi)(80\,\text{ft})(1.39 \times 10^{-3}\,\text{ft/min})} = 9.6\,\text{ft}$$

e. Determine the sidestream distance of the envelope far upstream of the pumping wells, Q/Bu:

$$\frac{Q}{Bu} = \frac{(50\,\text{gal/min})(1\,\text{ft}^3/7.48\,\text{gal})}{(80\,\text{ft})(1.39 \times 10^{-3}\,\text{ft/min})} = 60.0\,\text{ft}$$

f. The general shape of the envelope can be defined by using the above characteristic distances:

x (ft)	y (ft)	Note
0	9.6	Location of the first well
0	−9.6	Location of the second well
−9.6	0	Downstream distance (stagnation point)
0	30	Sidestream distance at the line of the wells
0	−30	Sidestream distance at the line of the wells
300*	60	Sidestream distance far upstream of the wells
300*	−60	Sidestream distance far upstream of the wells

* The sidestream distance far upstream of the wells, ±60 ft, should occur at $x = \infty$. A value of 300, which is ten times the sidestream distance at the line of wells, is used as the value of x.

Discussion

1. The sidestream distance at the line of the two pumping wells is twice that of the single well.
2. The sidestream distance far upstream of the two pumping wells is twice that of the single well.
3. The downstream distance of the two pumping wells is the same as that of the single pumping well. The calculated downstream distance,

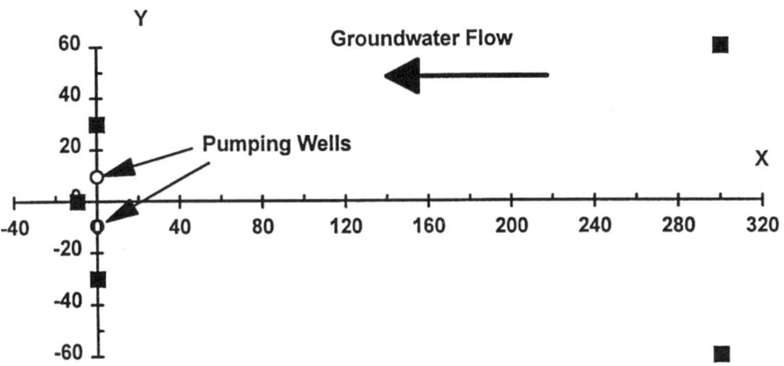

Figure E.VI.1.2C Capture zone of two wells.

$Q/2\pi Bu$, is along the x-axis. However, the affected distances directly downstream of these two wells should be slightly greater than $Q/2\pi Bu$.

Well spacing and number of wells

As mentioned earlier, it is important to determine the number of wells and their spacing in a groundwater remediation program. After the extent of the plume, and the direction and velocity of the groundwater flow have been determined, the following procedure can be used to determine the number of wells and their locations:

Step 1: Determine the groundwater pumping rate from aquifer testing or estimate the flow rate by using information of the aquifer materials.

Step 2: Draw the capture zone of one groundwater well (see Example VI.1.2A or VI.1.2B), using the same scale as the plume map.

Step 3: Superimpose the capture zone curve on the plume map. Make sure the direction of the groundwater of the capture zone curve matches that of the plume map.

Step 4: If the capture zone can completely encompass the extent of the plume, one pumping well is the optimum number. The location of the well on the capture zone curve is then copied to the plume map. One may want to reduce the groundwater extraction rate to have a smaller capture zone, but still sufficient to cover the entire plume.

Step 5: If the capture zone cannot encompass the entire extent of the plume, prepare the capture zone curves using two or more pumping wells until the capture zone can cover the entire plume. The locations of the wells on the capture zone curve are then copied to the plume map. (Note that the zones of influence of individual wells may overlap. Consequently, one may not be able to pump the same flow rate from each well in a network of wells as one can from a single well with the same allowable drawdown.)

Example VI.1.2D *Determine the number and locations of pumping wells for capturing a groundwater plume*

An aquifer (hydraulic conductivity = 1000 gpd/ft^2, gradient = 0.015, and aquifer thickness = 80 ft) is contaminated. The extent of the plume has been defined and it is shown in Figure E.VI.1.2D. (Each interval on the x-axis is 40 ft and that on the y-axis is 20 ft.)

Determine the number and locations of groundwater extraction wells for remediation. The design pumping rate of each well is 50 gpm.

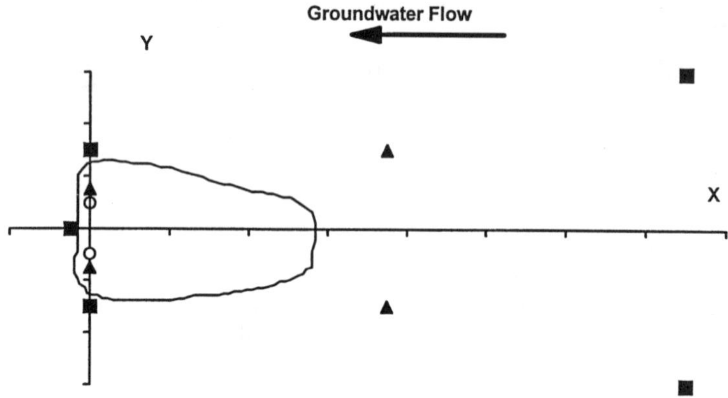

Figure E.VI.1.2D Capture zones of one and two wells.

Solution:

a. Plot the capture zone of a single well (same as Example VI.1.2B). The triangle symbols on the figure define the capture zone of this single well. As shown, this capture zone could not encompass the entire plume.
b. Plot the capture zone of two pumping wells (same as Example VI.1.2C). The square symbols on the figure define the capture zone of these two wells. As shown, this capture zone can encompass the entire plume. Consequently, using two pumping wells is optimum. The locations of these two pumping wells are shown as open circles in the figure.

VI.2 Above-ground groundwater treatment systems

VI.2.1 Activated carbon adsorption

Adsorption is the process that collects soluble substances in solution onto the surface of the adsorbent solids. Activated carbon is a universal adsorbent that adsorbs almost all types of organic compounds. Activated carbon particles have a large specific surface area. In activated carbon adsorption, the organics leave (or are removed from) the liquid by adsorbing onto the carbon surface. As the carbon bed becomes exhausted, as indicated by breakthrough of contaminants in the effluent, the carbon must be regenerated or replaced.

Common preliminary design of an activated carbon adsorption system includes sizing of the adsorber, determining the carbon-change (or regeneration) interval, and configuring the carbon units, when multiple carbon adsorbers are used.

Adsorption isotherm and adsorption capacity

In general, the amount of materials adsorbed depends on the characteristics of the solute and the activated carbon, the solute concentration, and the temperature. An adsorption isotherm describes the equilibrium relationship between the adsorbed solute concentration on the solid and the dissolved solute concentration in the bulk solution at a given temperature. The adsorption capacity of a given activated carbon for a specific compound is estimated from their isotherm data. The most commonly used adsorption models in environmental applications are the Langmuir and Freundlich isotherms, respectively:

$$q = \frac{abC}{1 + bC} \qquad \text{[Eq. VI.2.1]}$$

$$q = kC^n \qquad \text{[Eq. VI.2.2]}$$

where q is the adsorbed concentration (in mass of contaminant/mass of activated carbon), C is the liquid concentration (in mass of contaminant/volume of solution), and a, b, k, and n are constants. The adsorption concentration, q, obtained from Eq. VI.2.1 or VI.2.2. is the equilibrium value (the one in equilibrium with the liquid solute concentration). It should be considered as the theoretical adsorption capacity for a specified liquid concentration. The actual adsorption capacity in the field applications should be lower. Normally, design engineers take 25 to 50% of this theoretical value as the design adsorption capacity as a factor of safety. Therefore,

$$q_{actual} = (50\%)(q_{theoretical}) \qquad \text{[Eq. VI.2.3]}$$

The maximum amount of contaminants that can be removed or held ($M_{removal}$) by a given amount of activated carbon can be determined as

$$\begin{aligned} M_{removal} &= (q_{actual})(M_{carbon}) \\ &= (q_{actual})[(V_{carbon})(\rho_b)] \end{aligned} \qquad \text{[Eq. VI.2.4]}$$

where M_{carbon} is the mass, V_{carbon} is the volume, and ρ_b is the bulk density of activated carbon, respectively.

The following procedure can be used to determine the adsorption capacity of an activated carbon adsorber:

Step 1: Determine the theoretical adsorption capacity by using Eq. VI.2.1 or VI.2.2.
Step 2: Determine the actual adsorption capacity by using Eq. VI.2.3.
Step 3: Determine the amount of activated carbon in the adsorber.
Step 4: Determine the maximum amount of contaminants that can be held by the adsorber using Eq. VI.2.4.

Information needed for this calculation

- Adsorption isotherm
- Contaminant concentration of the influent liquid, C_{in}
- Volume of the activated carbon, V_{carbon}
- Bulk density of the activated carbon, ρ_b

Example VI.2.1A Determine the capacity of an activated carbon adsorber

Dewatering to lower the groundwater level for below-ground construction is often necessary. At a construction site, the contractor unexpectedly found that the extracted groundwater was contaminated with 5 mg/L toluene. The toluene concentration of the groundwater has to be reduced to below 100 ppb before discharge. To avoid further delay of the tight construction schedule, off-the-shelf 55-gal activated carbon units are proposed to treat the groundwater.

The activated carbon vendor provided the adsorption isotherm information. It follows the Langmuir model as q(kg toluene/kg carbon) = $[0.04C_e/(1 + 0.002C_e)]$, where C_e is in mg/L. The vendor also provided the following information regarding the adsorber:

Diameter of carbon packing bed in each 55-gal drum = 1.5 ft
Height of carbon packing bed in each 55-gal drum = 3 ft
Bulk density of the activated carbon = 30 lb/ft^3

Determine (a) the adsorption capacity of the activated carbon, (b) the amount of activated carbon in each 55-gal unit, and (c) the amount of the toluene that each unit can remove before exhausted.

Solution:

a. The theoretical adsorption capacity can be found by using Eq. VI.2.1 as

$$q(\text{kg/kg}) = \frac{0.004C_e}{1+0.002C_e} = \frac{(0.004)(5)}{1+(0.002)(5)} = 0.02\ \text{kg/kg}$$

The actual adsorption capacity can be found by using Eq. VI.2.3 as

$$q_{actual} = (50\%)q_{theoretical} = (50\%)(0.02) = 0.01 \text{ kg/kg}$$

b. Volume of the activated carbon inside a 55-gal drum = $(\pi r^2)(h)$

$$= (\pi)[(1.5/2)^2](3) = 5.3 \text{ ft}^3$$

Amount of the activated carbon inside a 55-gal drum = $(V)(\rho_b)$

$$= (5.3 \text{ ft}^3)(30 \text{ lb/ft}^3) = 159 \text{ lbs}$$

c. Amount of toluene that can be retained by a drum before the carbon becomes exhausted = (amount of the activated carbon)(actual adsorption capacity) = (159 lbs/drum)(0.01 lb toluene/lb activated carbon) = 1.59 lb toluene/drum.

Discussion

1. The bulk density of activated carbon is typically in the neighborhood of 30 lb/ft^3. The amount of activated carbon in a 55-gal drum is approximately 160 pounds.
2. The adsorption capacity of 0.01 kg/kg is equal to 0.01 lb/lb, or 0.01 g/g.
3. Care should be taken to use matching units for C and q in the isotherm equations.
4. The influent contaminant concentration in the liquid, not the effluent concentration, should be used in the isotherm equations to determine the adsorption capacity.

Empty bed contact time

To size the liquid-phase activated carbon system, the common criterion used in design is the empty bed contact time (EBCT). The typical EBCT ranges from 5 to 20 minutes, mainly depending on characteristics of the contaminants. Some compounds have a stronger tendency to adsorption, and the required EBCT would be shorter. Taking PCB and acetone as two extreme examples, PCB is very hydrophobic and will strongly adsorb to the activated carbon surface, while acetone is not readily adsorbable.

If the liquid flow rate (Q) is specified, the EBCT can be used to determine the required volume of the activated carbon adsorber (V_{carbon}) as

$$V_{carbon} = (Q)(EBCT) \qquad \text{[Eq. VI.2.5]}$$

Cross-sectional area

The typical hydraulic loading to carbon adsorbers is set to be 5 gpm/ft^2 or less. This parameter is used to determine the minimum required cross-sectional area of the adsorber (A_{carbon}):

$$A_{carbon} = \frac{Q}{Surface\ Loading\ Rate} \qquad \text{[Eq. VI.2.6]}$$

Height of the activated carbon adsorber

The required height of the activated carbon adsorber (H_{carbon}) can then be determined as

$$H_{carbon} = \frac{V_{carbon}}{A_{carbon}} \qquad \text{[Eq. VI.2.7]}$$

Contaminant removal rate by the activated carbon adsorber

The removal rate by a carbon adsorber ($R_{removal}$) can be calculated by using the following formula:

$$R_{removal} = (C_{in} - C_{out})Q \qquad \text{[Eq. VI.2.8]}$$

In practical applications, the effluent concentration (C_{out}) is kept below the discharge limit, which is often very low. Therefore, for a factor of safety, the term of C_{out} can be deleted from Eq. VI.2.8 in design. The mass removal rate is then the same as the mass loading rate ($R_{loading}$):

$$R_{removal} \sim R_{loading} = (C_{in})Q \qquad \text{[Eq. VI.2.9]}$$

Change-out (or regeneration) frequency

Once the activated carbon reaches its capacity, it should be regenerated or disposed of. The time interval between two regenerations or the expected service life of a fresh batch of activated carbon can be calculated by dividing the capacity of the activated carbon with the contaminant removal rate ($R_{removal}$) as

$$T = \frac{M_{removal}}{R_{removal}} \qquad \text{[Eq. VI.2.10]}$$

Configuration of the activated carbon adsorbers

If multiple activated carbon adsorbers are used, the adsorbers are often arranged in series and/or in parallel. If two adsorbers are arranged in series, the monitoring point can be located at the effluent of the first adsorber. A high effluent concentration from the first adsorber indicates that this adsorber is reaching its capacity. The first adsorber is then taken off-line, and the second adsorber is shifted to be the first adsorber. Consequently, the capacity of both adsorbers would be fully utilized and the compliance requirements are met. If there are two parallel streams of adsorbers, one stream can always be taken off-line for regeneration or maintenance and the continuous operation of the process is secured.

The following procedure can be used to complete the design of an activated carbon adsorption system:

Step 1: Determine the adsorption capacity as described earlier in this section (also see Ex. VI.2.1A).
Step 2: Determine the required volume of the activated carbon adsorber by using Eq. VI.2.5.
Step 3: Determine the required area of the activated carbon adsorber by using Eq. VI.2.6.
Step 4: Determine the required height of the activated carbon adsorber by using Eq. VI.2.7.
Step 5: Determine the contaminant removal rate or loading rate by using Eq. VI.2.9.
Step 6: Determine the amount of the contaminants that the carbon adsorber(s) can hold by using Eq. VI.2.4.
Step 7: Determine the service life of the activated carbon adsorber by using Eq. VI.2.10.
Step 8: Determine the optimal configuration when multiple adsorbers are used.

Information needed for this calculation

- Adsorption isotherm
- Contaminant concentration of the influent liquid, C_{in}
- Design hydraulic loading rate
- Design liquid flow rate, Q
- Bulk density of the activated carbon, ρ_b

Example VI.2.1B Design an activated carbon system for groundwater remediation

Dewatering to lower the groundwater level for below-ground construction is often necessary. At a construction site, the contractor unexpectedly found that the extracted groundwater was contaminated with 5 mg/L toluene. The toluene concentration of the groundwater has to be reduced to below 100 ppb before discharge. To avoid further delay of the tight construction schedule, off-the-shelf 55-gal activated carbon units are proposed to treat the groundwater. Use the following information to design an activated carbon treatment system (i.e., number of carbon units, configuration of flow, and carbon change-out frequency):

Wastewater flow rate = 30 gpm
Diameter of carbon packing bed in each 55-gal drum = 1.5 ft

Height of carbon packing bed in each 55-gal drum = 3 ft
Bulk density of GAC = 30 lb/ft^3
Adsorption isotherm: q(kg toluene/kg carbon) = $[0.04C_e/(1 + 0.002C_e)]$
where C_e is in mg/L

Solution:

a. The actual adsorption capacity has been found in Example VI.2.1A as 0.01 lb/lb.
b. Assuming an EBCT of 12 minutes, the required volume of the carbon adsorber can be found by using Eq. VI.2.5:

$$V_{carbon} = (Q)(EBCT) = [(30 \text{ gpm})(\text{ft}^3/7.48 \text{ gal})](12 \text{ min}) = 48.1 \text{ ft}^3$$

c. Assuming a design hydraulic loading of 5 gpm/ft^2 or less, the required cross-sectional area for the carbon adsorption can be found by using Eq. VI.2.6:

$$A_{carbon} = \frac{Q}{Surface\,Loading\,Rate} = \frac{30\,\text{gpm}}{5\,\text{gpm/ft}^2} = 6 \text{ ft}^2 \quad \text{[Eq. VI.2.6]}$$

d. If the adsorption system is tailor-made, then a system with a cross-sectional area of 6 ft^2 and a height of 8 ft (= 48.1/6) will do the job. However, if the off-the-shelf 55-gal drums are to be used, we need to determine the number of drums that will provide the required cross-sectional area.
Area of the activated carbon inside a 55-gal drum = $(\pi r^2) = (\pi)[(1.5/2)^2]$ = 1.77 ft^2/drum.
Number of drums in parallel to meet the required hydraulic loading rate = (6 ft^2) ÷ (1.77 ft^2/drum) = 3.4 drums.
So, use four drums in parallel. The total cross-sectional area of four drums is equal to 7.08 ft^2 (= 1.77 × 4).
e. The required height of the activated carbon adsorber can be found by using Eq. VI.2.7:

$$H_{carbon} = \frac{V_{carbon}}{A_{carbon}} = \frac{48.1}{7.08} = 6.8 \text{ ft}$$

The height of activated carbon in each drum is 3 ft. The number of drums required in series to meet the required height of 6.8 ft can be found as

Number of drums in-series to meet the required height
= (6.8 ft) ÷ (3 ft/drum) = 2.3 drums

So, use three drums in series. The total volume of activated carbon in twelve drums is equal to 63.6 ft^3 (= 5.3 × 4 × 3).

f. Determine the contaminant removal rate or loading rate by using Eq. VI.2.9:

$$R_{removal} \sim R_{loading} = (C_{in})Q = (5 \text{ mg/L})$$
$$[(30 \text{ gal/min})(3.785 \text{ L/gal}) \times (1440 \text{ min/d})]$$
$$= 817{,}560 \text{ mg/d} = 1.8 \text{ lb/d}$$

g. Determine the amount of the contaminants that the carbon adsorber(s) can hold by using Eq. VI.2.4:

$$M_{removal} = (q_{actual})[(V_{carbon})(\rho_b)] = (0.01)[(63.6)(30)] = 19.1 \text{ lb}$$

h. Determine the service life of the carbon adsorbers by using Eq. VI.2.10:

$$T = \frac{M_{removal}}{R_{removal}} = \frac{19.1 \text{ lb}}{1.8 \text{ lb/d}} = 10.6 \text{ days}$$

Discussion

1. The configuration is 4 drums in parallel and 3 drums in series (a total of 12 drums). Care should be taken to minimize the head loss due to numerous piping connections.
2. A 55-gal activated carbon drum normally costs several hundred dollars. In this example, 12 drums last less than 11 days. The disposal or regeneration cost should also be added, and it makes this option relatively expensive. If a long-term treatment is needed, one may want to switch to larger activated carbon adsorbers or to other treatment methods.

VI.2.2 Air stripping

Air stripping is a physical process that enhances volatilization of organic compounds from water by passing clean air through it. It is one of the commonly used processes for treating groundwater contaminated with VOCs.

An air stripping system creates air and water interfaces to enhance mass transfer between the air and liquid phases. Although there are several system configurations commercially available, including tray columns, spray systems, diffused aeration, and packed columns (or packed towers), use of packed towers is the most popular alternative for groundwater remediation applications.

Process description

In a packed-column air stripping tower, the air and the contaminated groundwater streams flow countercurrently through a packing column. The packing provides a large surface area for VOCs to migrate from the liquid stream to the air stream. A mass balance equation can be derived by letting the amount of contaminants removed from the liquid be equal to the amount of the contaminants entering the air:

$$Q_w(C_{in} - C_{out}) = Q_a(G_{out} - G_{in}) \qquad \text{[Eq. VI.2.11]}$$

where C = contaminant concentration in the liquid phase (mg/L), G = contaminant concentration in the air phase (mg/L), Q_a = air flow rate (L/min), and Q_w = liquid flow rate (L/min).

For an ideal case where the influent air contains no contaminants (G_{in} = 0) and the groundwater is completely decontaminated (C_{out} = 0), Eq. VI.2.11 can be simplified as

$$Q_w(C_{in}) = Q_a(G_{out}) \qquad \text{[Eq. VI.2.12]}$$

Assume that Henry's law applies and the effluent air is in equilibrium with the influent water, then

$$G_{out} = H^*C_{in} \qquad \text{[Eq. VI.2.13]}$$

where H^* is Henry's constant of the compound of concern in a dimensionless form.

Combining Eqs. VI.2.12 and VI.2.13, the following relationship can be developed:

$$H^*\left(\frac{Q_a}{Q_w}\right)_{\min} = 1 \qquad \text{[Eq. VI.2.14]}$$

The $(Q_a/Q_w)_{min}$ is the minimum air-to-water ratio (in vol/vol), and this is the air-to-water ratio for the above-mentioned ideal case. The actual air-to-water ratio is often chosen to be a few times larger than the minimum air-to-water ratio.

The stripping factor (S), which is the product of the dimensionless Henry's constant and the air-to-water ratio, is commonly used in air stripping design:

$$S = H^*\left(\frac{Q_a}{Q_w}\right) \qquad \text{[Eq. VI.2.15]}$$

The stripping factor is equal to unity for the above-mentioned ideal case. It would require a packing height of infinity to achieve the perfect removal. For field applications, the values of S should be greater than one.

Practical values of S range from 2 to 10. Operating a system with values of S larger than 10 may not be economical. In addition, a high air-to-water ratio may cause an unfavorable phenomenon, called "flooding," in an air stripping operation.

The following procedure can be used to determine the air flow rate for a given liquid flow rate:

Step 1: Convert Henry's constant to its dimensionless value using the formula given in Table II.3.B.

Step 2: If the stripping factor is known or selected, determine the air-to-water ratio by using Eq. VI.2.15. Go to Step 4.

Step 3: If the stripping factor is not known or selected, determine the minimum air-to-water ratio by using Eq. VI.2.14. Obtain the design air-to-water ratio by multiplying the minimum air-to-water ratio with a value between 2 and 10. Go to Step 4.

Step 4: Determine the required air flow rate by multiplying the liquid flow rate with the air-to-water ratio determined from Step 2 or Step 3.

Information needed for this calculation

- Henry's constant
- Stripping factor, S
- Design liquid flow rate, Q

Example VI.2.2A *Determine the air-to-water ratio of an air stripper*

A packed-column air stripper is designed to reduce the chloroform concentration in the extracted groundwater. The concentration is to be reduced from 50 mg/L to 0.05 mg/L (50 ppb). Determine (1) the minimum air-to-water ratio, (2) the design air-to-water ratio, and (3) the design air flow rate. Use the following information in calculations:

Henry's constant for chloroform = 128 atm
Stripping factor = 3
Temperature of the water = 15°C
Extracted groundwater flow rate = 120 gpm

Solution:

a. Use the formula in Table II.3.B to convert the Henry's constant to its dimensionless value:

$$H = \frac{H^*RT(1000\gamma)}{W} = \frac{H^*(0.082)(273+15)(1000)(1)}{18} = 128$$

So,

$$H^* = 0.098 \text{ (dimensionless)}$$

b. Use Eq. VI.2.14 to determine the minimum air-to-water-ratio:

$$H^*\left(\frac{Q_a}{Q_w}\right)_{\min} = 1 = (0.098)\left(\frac{Q_a}{Q_w}\right)_{\min}$$

So,

$$(Q_a/Q_w)_{min} = 10.25 \text{ (dimensionless)}$$

c. Use Eq. VI.2.15 to determine the air-to-water-ratio:

$$H^*\left(\frac{Q_a}{Q_w}\right) = S = 3 = (0.098)\left(\frac{Q_a}{Q_w}\right)$$

So,

$$(Q_a/Q_w) = 30.75 \text{ (dimensionless)}$$

d. Determine the required air flow rate by multiplying the liquid flow rate with the air-to-water ratio:

$$Q_a = Q_w \times (Q_a/Q_w) = (120 \text{ gpm})(30.75) = 3690 \text{ gal/min} = 493 \text{ ft}^3\text{/min}$$

Discussion. A stripping factor of three means the ratio of the design and the minimum air-to-water ratio are three. The design air-to-water ratio can be obtained by multiplying the minimum air-to-water ratio with the stripping factor.

Column diameter

One of the key components in sizing an air stripper is to determine the diameter of the column. The diameter depends mainly on the liquid flow rate. The higher the liquid flow rate is, the larger the column diameter should be. Typical liquid hydraulic loading rate to an air stripping column is kept to 20 gpm/ft^2 or less. This parameter is often used to determine the required cross-sectional area of the stripping column ($A_{stripping}$):

$$A_{stripping} = \frac{Q}{Surface\ Loading\ Rate} \qquad \text{[Eq. VI.2.16]}$$

Packing height

The required depth of the packing column (Z) for a specific removal efficiency is another important design component. A taller column would be required to achieve a higher removal efficiency. The packing height can be conveniently determined using the transfer unit concept:

$$Z = (HTU) \times (NTU) \qquad \text{[Eq. VI.2.17]}$$

where *HTU* is the height of transfer unit and *NTU* is the number of transfer unit.

The *HTU* depends heavily on the hydraulic loading rate and the overall mass transfer coefficient, K_La. [Note: K_L is the rate constant (m/sec) and "*a*" is the specific surface area (m^2/m^3). K_La has a unit of 1/time.] The K_La value for a specific application can be best determined from pilot testing, and there are also empirical equations available to estimate the value of K_La. Values of K_La in common air stripping columns used in groundwater remediation range from 0.01 to 0.05 sec^{-1}. HTU has a unit of length and can be determined as

$$HTU = \frac{L}{(K_La)} \qquad \text{[Eq. VI.2.18]}$$

where *L* is the liquid hydraulic loading rate in length/time.

The *NTU* can be determined by using the following formula:

$$NTU = \left(\frac{S}{S-1}\right)\ln\left\{\frac{(C_{in} - G_{in}/H^*)}{(C_{out} - G_{in}/H^*)}\left[\frac{S-1}{S}\right] + \frac{1}{S}\right\}$$

$$= \left(\frac{S}{S-1}\right)\ln\left\{\left(\frac{C_{in}}{C_{out}}\right)\left[\frac{S-1}{S}\right] + \frac{1}{S}\right\} \quad (\text{for } G_{in} = 0) \qquad \text{[Eq. VI.2.19]}$$

where *S* is the stripping factor, H^* is Henry's constant in dimensionless form, *C* is the contaminant concentration in liquid, and *G* is the contaminant concentration in air.

The following procedure can be used to size an air stripping column:

Step 1: Determine the required cross-sectional area of the air stripper by using Eq. VI.2.16. Then, determine the diameter of the column corresponding to this calculated area. Round up the diameter value to the next half or whole foot.

Step 2: Use the newly found diameter to calculate the cross-sectional area and then the hydraulic loading rate. Use Eq. VI.2.18 to find the *HTU*.

Step 3: Determine the stripping factor, if not known or specified, by using Eq. VI.2.15.
Step 4: Use Eq. VI.2.19 to find the *NTU*.
Step 5: Use Eq. VI.2.17 to find the packing height, Z.

Information needed for this calculation

- Henry's constant
- Stripping factor, *S*
- Design hydraulic loading rate
- Design liquid flow rate, *Q*
- Overall mass transfer coefficient, K_La
- Influent contaminant concentration in liquid, C_{in}
- Effluent contaminant concentration in liquid, C_{out}
- Contaminant concentration in influent air, G_{in}

Example VI.2.2B *Sizing an air stripper for groundwater remediation*

A packed-column air stripper is designed to reduce the chloroform concentration in the extracted groundwater. The concentration is to be reduced from 50 mg/L to 0.05 mg/L (50 ppb). Size the air stripper by determining cross-sectional surface area, packing height, and air flow rate.

Use the following information in the calculations:

Henry's constant for chloroform = 128 atm
Stripping factor = 3
Temperature of the water = 15°C
Extracted groundwater = 120 gpm
K_La = 0.01/s
Packing = Jaeger 3″ Tri-packs
Hydraulic loading rate = 20 gpm/ft^2
Contaminant concentration in the influent air = 0

Solution:

a. As shown in Example VI.2.2B, the dimensionless value of Henry's constant is equal to 0.098, and the air flow rate is determined to be 493 ft^3/min.
b. Use Eq. VI.2.14 to determine the required cross-sectional area:

$$A_{stripping} = \frac{Q}{Surface\ Loading\ Rate} = \frac{120\,\text{gpm}}{20\,\text{gpm/ft}^2} = 6\ \text{ft}^2$$

Diameter of the air stripping column = $(4A/\pi)^{1/2} = (4 \times 6/\pi)^{1/2} = 2.76$ ft. So, d = 3 ft.

c. Use this newly found diameter to find the hydraulic loading rate: Cross-sectional area of the column = $\pi d^2/4 = \pi(3)^2/4 = 7.1$ ft^2. Hydraulic loading rate to the column = Q/A = [(120 gpm)(ft^3/7.48 gal)] ÷ 7.1 ft^2 = 2.26 ft/min = 0.0377 ft/s.

d. Use Eq. VI.2.18 to determine the *HTU* value:

$$HTU = \frac{L}{(K_L a)} = \frac{0.0377\,\text{ft/s}}{0.01/\text{s}} = 3.77\,\text{ft}$$

e. Use Eq. VI.2.19 to determine the *NTU* value:

$$NTU = \left(\frac{S}{S-1}\right)\ln\left\{\left(\frac{C_{in}}{C_{out}}\right)\left[\frac{S-1}{S}\right]+\frac{1}{S}\right\} \quad (\text{for } G_{in} = 0)$$

$$= \left(\frac{3}{3-1}\right)\ln\left\{\left(\frac{50}{0.05}\right)\left[\frac{3-1}{3}\right]+\frac{1}{3}\right\} = 9.75$$

f. Use Eq. VI.2.17 to determine the packing height:

$$Z = (HTU) \times (NTU) = (3.77\text{ ft})(9.75) = 36.8\text{ ft}$$

Discussion

1. The typical hydraulic loading rate, 20 gpm/ft^2, is much larger than that for the activated carbon adsorbers, 5 gpm/ft^2.
2. The required packing height of 36.8 ft will make the total height of the air stripper well over 40 ft. This may not be acceptable in most project locations. If this is the case, one may consider having two shorter air strippers in series.

VI.2.3 Advanced oxidation process

Advanced oxidation process (AOP) refers to an oxidation process assisted by ultraviolet (UV) irradiation. In AOP, high-power lamps emit UV radiation through quartz sleeves into contaminated water. An oxidizing agent, typically hydrogen peroxide, ozone, or a combination of these two, is added. The oxidizing agent is activated by the UV light to form hydroxyl radicals, which have very strong oxidizing power. These radicals destroy the organic contaminants in water.

Reactor sizing

In a typical AOP, oxidizing reagents are often injected and mixed using metering pumps and in-line static mixers. The groundwater then flows sequentially through one or more UV reactors. The reactors are often considered as plug-flow type, and the reactions follow first-order kinetics. Eq.

IV.3.10 describes the relationship among the influent concentration, effluent concentration, retention time, and reaction rate constant for plug flow reactors. It is repeated here for the AOP reactors as

$$\frac{C_{out}}{C_{in}} = e^{-k(V/Q)} = e^{-k\tau} \qquad \text{[Eq. VI.2.20]}$$

where C is the contaminant concentration in groundwater, V is the reactor volume, Q is the groundwater flow rate, k is the rate constant, and τ is the hydraulic retention time.

Example VI.2.3 Sizing the reactor for an advanced oxidation process

UV/ozone treatment is selected to remove TCE from an extracted groundwater stream (TCE concentration = 400 ppb). A pilot study was conducted and found that, with a hydraulic retention time of 2 minutes, the system could reduce TCE concentration from 400 to 16 ppb. However, the discharge limit for TCE is 3.2 ppb. Assuming the reactors are of ideal plug flow type and the reaction is first-order, how many reactors would you recommend using?

Solution:

a. Use Eq. VI.2.20 to determine the reaction rate constant:

$$\frac{C_{out}}{C_{in}} = \exp[-k(\tau)] = \frac{16}{400} = \exp[-2k]$$

So,

$$k = 1.61/\text{min}$$

b. Use Eq. VI.2.20 again to determine the required retention time to reduce the TCE concentration below the discharge limit:

$$\frac{C_{out}}{C_{in}} = \exp[-k(\tau)] = \frac{3.2}{400} = \exp[-1.61(\tau)]$$

$$\tau = 3.0 \text{ minutes}$$

Thus, it requires two reactors.

c. Use Eq. VI.2.20 again to determine the final effluent TCE concentration (τ = 4 minutes because two reactors were used):

$$\frac{C_{out}}{C_{in}} = \exp[-k(\tau)] = \frac{C_{out}}{400} = \exp[-1.61(4)]$$

$$C_{out} = 0.64 \text{ ppb.}$$

Discussion

1. For PFRs, the final concentration from two identical reactors in series is the same as that from two identical reactors in parallel.
2. A pilot–scale test to determine the removal efficiency and the reaction rate constant is always recommended for AOPs.
3. Example IV.5.1D is another example for AOPs.

VI.2.4 Metal removal by precipitation

Elevated heavy metal concentrations may occur in extracted groundwater or in liquid waste streams. Chemical precipitation is a common removal method for inorganic heavy metals in groundwater or wastewater. The hydroxides of heavy metals are formed at high pH and are usually insoluble. Lime or caustic soda is often added to precipitate the metals. The solubility of metal hydroxides is sensitive to pH, and the reaction can be expressed in a general form:

$$M(OH)_n \Leftrightarrow M^{n+} + nOH^- \qquad \text{[Eq. VI.2.21]}$$

where M represents the heavy metal, OH^- is the hydroxide ion, and n is the valence of the metal.

The equilibrium equation can be written as

$$K_{sp} = [M^{n+}][OH^-]^n \qquad \text{[Eq. VI.2.22]}$$

where K_{sp} is the equilibrium constant (often called the solubility product), $[M^{n+}]$ is the molar concentration of the heavy metal, and $[OH^-]$ is the molar concentration of hydroxide. For example, the K_{sp} values for $Cr(OH)_3$, $Fe(OH)_3$, and $Mg(OH)_2$ at 25°C are 6×10^{-31} M^4, 6×10^{-36} M^4, and 9×10^{-12} M^3, respectively.

Example VI.2.4 Chemical precipitation for magnesium removal

Sodium hydroxide is added to a CFSTR to remove magnesium ion from an extracted groundwater stream ($Q = 150$ gpm). The temperature of the reactor is kept at 25°C and pH = 11. The influent Mg^{2+} concentration is 100 mg/L. If the solids are settled to 1% by weight, estimate

a. The Mg^{2+} in the treated effluent (mg/L)
b. Rate of $Mg(OH)_2$ produced (lb/day)
c. Rate of sludge produced (lb sludge/day)

(Note: The solubility product of $Mg(OH)_2$ is 9×10^{-12} M^3 at 25°C; molecular weight of Mg = 24.3.)

Solution:

a. Write the reaction of precipitation first:

$$Mg(OH)_2 \Leftrightarrow Mg^{2+} + 2OH^-$$

At pH =11, the hydroxide concentration $[OH^-]$ is equal to 10^{-3} M. Use the solubility product equation to detemine the magnesium concentration as

$$K_{sp} = [Mg^{2+}][OH^-]^2 = 9 \times 10^{-12} = [Mg^{2+}][10^{-3}]^2$$

$$[Mg^{2+}] = 9 \times 10^{-6}\ M = (9 \times 10^{-6}\ mole/L)(24.3\ g/mole)$$
$$= 2.19 \times 10^{-4}\ g\ /L = 0.22\ mg/L$$

b. As shown in (a), 1 mole of $Mg(OH)_2$ formed for each mole of Mg^{2+} removed. Since the molecular weight of $Mg(OH)_2$ is equal to 58.3, the rate of $Mg(OH)_2$ produced can be found as:

$$\text{Rate of } Mg(OH)_2 \text{ produced} = (\text{rate of } Mg^{2+} \text{ removed})(58.3/24.3)$$
$$= \{[Mg^{2+}]_{in} - [Mg^{2+}]_{out}\}Q \times (58.3/24.3)$$
$$= [(100 - 0.22)\ mg/L][(150\ gpm)(3.785\ L/gal)](58.3/24.3)$$
$$= 136{,}000\ mg/min = 136\ g/min = 431\ lb/day$$

c. Since the solids are settled to 1% by weight, the rate of sludge production can be found as

$$\text{Rate of sludge produced} = \text{Rate of } Mg(OH)_2 \text{ produced} \div 1\%$$
$$= 431\ lb/day \div 1\% = 43{,}100\ lb/day$$

VI.2.5 Biological treatment

Above-ground biological reactors are also used to remove organics from contaminated groundwater. In general, the bioreactors for removing dissolved organics from water or wastewater can be classified into two types: suspended growth or attached growth. The most common suspended growth type is the activated sludge process, while that for the attached growth type is the trickling filter process.

Biological systems used in groundwater remediation are usually much smaller in scale compared to those in most municipal or industrial wastewater treatment plants. The reactors often consist of packing material to support the bacterial growth and are similar to the attached-growth bioreactors in principle. Since the biological process is relatively complicated and affected by many factors, a pilot study is usually recommended to predict the performances of the biological systems. For the trickling filter type of bioreactors, the following empirical equation is often used:[4]

$$\frac{C_{out}}{C_{in}} = \exp[-kD(Q/A)^{-0.5}] \qquad \text{[Eq. VI.2.23]}$$

where C_{out} = contaminant concentration in the reactor effluent, mg/L, C_{in} = contaminant concentration in the reactor influent, mg/L, k = rate constant corresponding to a packing depth of D, $(\text{gpm})^{0.5}/\text{ft}$, D = depth of the filter, ft, Q = liquid flow rate, gpm, and A = cross-sectional area of the packing material, ft^2.

The hydraulic loading rate to a bioreactor is often small at 0.5 gpm/ft^2 or less. If the hydraulic loading rate is known, the following equation can be used to determine the cross-sectional area of the bioreactor:

$$A_{bioreactor} = \frac{Q}{Surface\ Loading\ Rate} \qquad \text{[Eq. VI.2.24]}$$

When a rate constant determined from one packing depth is used to design a bioreactor of a different packing depth, the following empirical formula should be used to adjust the rate constant:

$$k_2 = k_1\left(\frac{D_1}{D_2}\right)^{0.3} \qquad \text{[Eq. VI.2.25]}$$

where k_1 = rate constant corresponds to a filter of depth D_1, k_2 = rate constant corresponds to a filter of depth D_2, D_1 = depth of filter #1, and D_2 = depth of filter #2.

The following procedure can be used to size an attached-growth bioreactor:

Step 1: Select a desirable packing height, D. Adjust the rate constant to the selected packing height, if necessary, by using Eq. VI.2.25.

Step 2: Determine the hydraulic loading rate of the bioreactor by using Eq. VI.2.23.

Step 3: Determine the required cross-sectional area by using Eq. VI.2.24. Calculate the diameter of the bioreactor corresponding to this area. Round up the diameter value to the next half or whole ft.

If the calculated cross-sectional area is too large, select a larger packing depth and restart from Step 1.

Information needed for this calculation

- Rate constant, k
- Influent contaminant concentration, C_{in}
- Effluent contaminant concentration, C_{out}
- Design liquid flow rate, Q

Example VI.2.5A Sizing an above-ground bioreactor for groundwater remediation

A packed-bed bioreactor is designed to reduce the toluene concentration in the extracted groundwater. The concentration is to be reduced from 4 mg/L to 0.1 mg/L (100 ppb). The packing depth has been selected as 3 ft. Determine the required diameter of the bioreactor.

Use the following information in the calculations:

Rate constant = 0.9 $(\text{gpm})^{0.5}$/ft at 20°C (for 2 ft packing depth)
Temperature of the water = 20°C
Groundwater extraction rate = 20 gpm

Solution:

a. Use Eq. VI.2.25 to adjust the rate constant:

$$k_2 = k_1\left(\frac{D_1}{D_2}\right)^{0.3} = (0.9)\left(\frac{2}{3}\right)^{0.3} = 0.80\ (\text{gpm})^{0.5}/\text{ft}$$

b. Use Eq. VI.2.23 to determine the surface loading rate, Q/A:

$$\frac{C_{out}}{C_{in}} = \exp[-kD(Q/A)^{-0.5}] = \frac{0.1}{4} = \exp[-(0.80)(3)(Q/A)^{-0.5}]$$

$$Q/A = 0.423\ \text{gpm/ft}^2$$

c. Use Eq. VI.2.24 to determine the required cross-sectional area:

$$A_{bioreactor} = \frac{Q}{\textit{Surface Loading Rate}} = \frac{20\ \text{gpm}}{0.423\ \text{gpm/ft}^2} = 47.2\ \text{ft}^2$$

Diameter of the bioreactor packing = $(4A/\pi)^{1/2} = (4 \times 47.2/\pi)^{1/2} = 7.76$ ft. So,

$$d = 8\ \text{ft}$$

d. Assuming the packing material only occupies a small fraction of the total reactor volume, the hydraulic retention time can be estimated by hydraulic retention time = $(V/Q) = (Ah)/Q = (47.2 \text{ ft}^2)(3 \text{ ft}) \div [20 \text{ gpm } (1 \text{ ft}^3/7.48 \text{ gal})] = 53$ minutes.

Discussion. It is relatively difficult for the effluent of the bioreactors to meet the ppb level of the discharge requirements. Activated carbon adsorbers are often used as the polishers to treat the bioreactors' effluent before discharge.

VI.3 In situ groundwater remediation

VI.3.1 In situ bioremediation

Biological in situ treatment of organic contaminants in aquifers is usually accomplished by enhancing activities of indigenous subsurface microorganisms. Most of the in situ bioremediation is practiced in the aerobic mode. The microbial activities are enhanced by addition of inorganic nutrients and oxygen into the groundwater plume. The typical process consists of withdrawal of groundwater, addition of oxygen and nutrients, reinjection of the enriched groundwater through injection wells, or infiltration galleries.

Oxygen supply

Groundwater naturally contains low concentrations of oxygen. Even if it is fully saturated with air, the saturated dissolved oxygen (DO) concentration in groundwater would only be in the neighborhood of 9 mg/L at 20°C. Water saturated with pure oxygen would have a DO concentration five times higher, at approximately 45 mg/L. In most cases, oxygen addition through air- or oxygen-saturated water cannot meet the oxygen demand for biodegradation of contaminants in the groundwater plume. This explains why hydrogen peroxide is commonly used as the source of oxygen for in situ groundwater bioremediation. As much as 500 mg/L of oxygen can be supplied through hydrogen peroxide addition. Higher concentrations of hydrogen peroxide can be added, but the water may become toxic to microorganisms. Each mole of hydrogen peroxide in water will dissociate into one mole oxygen and two moles of water as

$$2H_2O_2 \Rightarrow 2H_2O + O_2 \qquad \text{[Eq. VI.3.1]}$$

Example VI.3.1A Determine the necessity of oxygen addition for in situ groundwater bioremediation

A subsurface is contaminated with gasoline. The average dissolved gasoline concentration of the groundwater samples is 20 mg/L. In situ bioremediation is being considered for aquifer restoration. The aquifer has the following characteristics:

Porosity = 0.35
Organic content = 0.02
Subsurface temperature = 20°C
Bulk density of aquifer materials = 1.8 g/cm^3
DO concentration in the aquifer = 4.0 mg/L

Illustrate that the addition of oxygen to the aquifer is necessary to support biodegradation of the intruding gasoline contaminants.

Strategy. The gasoline in the saturated zone will be dissolved in the groundwater or adsorbed onto the surface of the aquifer materials (assuming the free-product phase is absent). Since only the contaminant concentration in the groundwater is known, we have to estimate the amount of gasoline adsorbed on the soil by using the partition equation discussed earlier in Chapter two. In addition, the physicochemical data for gasoline are often not available because gasoline is a mixture of compounds. We will use the data of one of the common components in gasoline, such as toluene, when the data for gasoline are not available.

Solution:
Basis = 1 m^3 of aquifer.

a. From Table II.3.C, the following physicochemical property of toluene was obtained:

$$\text{Log } K_{ow} = 2.73$$

Use Eq. II.3.14 to find K_{oc}:

$$K_{oc} = 0.63K_{ow} = 0.63\,(10^{2.73}) = (0.63)(537) = 338$$

Use Eq. II.3.12 to find K_p:

$$K_p = f_{oc}K_{oc} = (0.02)\,(338) = 6.8 \text{ L/kg}$$

Use Eq. II.3.11 to find the contaminant concentration adsorbed onto the solid:

$$X = K_pC = (6.8 \text{ L/kg})\,(20 \text{ mg/L}) = 136 \text{ mg/kg}$$

b. Determine the total mass of the contaminant present in the aquifer (per m^3).
Mass of the aquifer matrix = (1 m^3)(1800 kg/m^3) = 1800 kg.
Mass of the contaminant adsorbed on the solid surface = $(S)(M_s)$ = (136 mg/kg)(1800 kg) = 244,800 mg = 245 g.

Void space of the aquifer = $V\phi$ = (1 m^3)(35%)= 0.35 m^3 = 350 L.
Mass of the contaminant dissolved in the groundwater = (C)(V_l) = (20 mg/L)(350 L) = 7000 mg = 7.0 g.
Total mass of the contaminant in the aquifer = dissolved + adsorbed = 7 + 245 = 252 grams of gasoline/m^3 of aquifer.

c. The amount of oxygen present in the groundwater = (V_l)(DO) = (350 L)(4 mg/L) = 1080 mg = 1.08 g.
d. Use the 3.08 ratio to determine the oxygen requirements for complete oxidation (see Section V.2.4 for details):

Oxygen requirement = (3.08)(252) = 779 grams >> 1.08 g

As demonstrated, the oxygen contained in the groundwater of the aquifer is negligible when compared to the amount of oxygen required for complete aerobic biodegradation.

e. If the groundwater is brought to the surface and aerated with air, the saturated dissolved oxygen concentration in water at 20°C is approximately 9 mg/L. When this groundwater is recharged back to the contaminated zone, the maximum amount of additional oxygen added to the aquifer per pore volume can be found as

The amount of oxygen added by water saturated with air
= (V_l)(DO_{sat}) = (350 L)(9 mg/L) = 3150 mg = 3.15 g

Amount of oxygen-enriched water needed to meet the oxygen demand (expressed as the number of pore volumes of the plume) = (779/3.15) = 247.

Discussion

1. As shown in (e), the plume has to be flushed 247 times with air-saturated water to meet the oxygen requirement.
2. If the extracted water is aerated with pure oxygen, the saturated DO will be five times higher and the required flushing will be five times less.
3. Fraction of the contaminant in the dissolved phase = (mass of contaminant in the dissolved phase)/(total contaminant mass in the aquifer) = (7.0)/(7.0 + 245) = 2.8%. It shows that the contaminant in the pore liquid only accounts for a small portion of the total contaminant mass in the aquifer.

Example VI.3.1B *Determine the effectiveness of hydrogen peroxide addition as an oxygen source for bioremediation*

As illustrated in Example VI.3.1A, it would take a tremendous amount of water, whether it is saturated with air or pure oxygen, to meet the oxygen demand for in situ groundwater bioremediation. Addition of hydrogen peroxide becomes a popular alternative. Because of the biocidal potential of hydrogen peroxide, the maximum hydrogen peroxide in water is often kept below 1000 mg/L for bioremediation applications. Determine the amount of oxygen that 1000 mg/L of hydrogen peroxide can provide.

Solution:

a. From Eq. VI.3.1, one mole of hydrogen peroxide can yield a half mole of oxygen:

$$2H_2O_2 \rightarrow 2H_2O + O_2$$

Molecular weight of hydrogen peroxide = (1 × 2) + (16 × 2) = 34.
Molecular weight of oxygen = 16 × 2 = 32.

b. Molar concentration of 1000 mg/L hydrogen peroxide = (1000 mg/L) ÷ (34,000 mg/mole) = 29.4×10^{-3} mole/L.
Molar concentration of oxygen (assume 100% dissociation of hydrogen peroxide) = 29.4×10^{-3} mole/L ÷ 2 = 14.7×10^{-3} mole/L.
Mass concentration of oxygen in water from hydrogen peroxide addition = (14.7×10^{-3} mole/L) × 32 g/mole = 470 mg/L.

Nutrient addition

Nutrients for microbial activity usually exist in the subsurface. However, with the presence of organic contaminants, additional nutrients are often needed to support the bioremediation. The nutrients to enhance microbial growth are assessed primarily on the nitrogen and phosphorus requirements of the microorganisms. The suggested C:N:P molar ratio is 120:10:1, as shown in Table V.2.A. The nutrients are typically added at concentrations ranging from 0.05 to 0.02% by weight.[1]

Example VI.3.1C Determine the nutrient requirement for in situ groundwater bioremediation

A subsurface is contaminated with gasoline. The average dissolved gasoline concentration of the groundwater samples is 20 mg/L. In situ bioremediation is being considered for aquifer restoration. The aquifer has the following characteristics:

Porosity = 0.35
Organic content = 0.02
Subsurface temperature = 20°C
Bulk density of aquifer materials = 1.8 g/cm^3

Assuming no nutrients are available in the groundwater for bioremediation and the optimal molar C:N:P ratio has been determined as 100:10:1, determine the amount of nutrients needed to support the biodegradation of the contaminants. If the plume is to be flushed with 100 pore volumes of oxygen- and nurient-enriched water, what would be the required nutrient concentration of this reinjected water?

Solution:

Basis = 1 m^3 of aquifer.

a. From Example VI.3.1A, the total mass of contaminants in the aquifer = 252 g/m^3.
b. Assume that the gasoline has a formula the same as heptane, C_7H_{16}. Molecular weight of gasoline = 7 × 12 + 1 × 16 = 100 and moles of gasoline = 252/100 = 2.52 g-mole.
c. Determine the number of moles of C. Since there are 7 carbon atoms in each gasoline molecule, as indicated by its formula, C_7H_{16}, then

$$\text{Moles of C} = (2.52)(7) = 17.7 \text{ g-mole}$$

d. Determine the number of moles of N needed (using the C:N:P ratio of 100:10:1).
Moles of N needed = (10/100)(17.7) = 1.77 g-mole.
Amount of nitrogen needed = 1.77 × 14 = 24.8 g/m^3 of aquifer.
Moles of $(NH_4)_2SO_4$ needed = 1.77 ÷ 2 = 0.885 g-mole (each mole of ammonium sulfate contains two moles of N).
Amount of $(NH_4)_2SO_4$ needed = (0.885)[(14 + 4)(2) + 32 + (16)(4)] = 117 g/m^3 of aquifer.
e. Determine the number of moles of P needed (using the C:N:P ratio of 100:10:1).
Moles of P needed = (1/100)(17.7) = 0.177 g-mole.
Moles of $Na_3PO_4 \cdot 12H_2O$ needed = 0.177 g-mole.
Amount of phosphorus needed = 0.177 × 31 = 5.5 g/m^3 of aquifer.
Amount of $Na_3PO_4 \cdot 12H_2O$ needed = (0.177)[(23)(3) + 31 + (16)(4)+ (12)(18)] = 67 g/m^3 of aquifer.
f. The total nutrient requirement = 117 + 67 = 184 g/m^3 of aquifer.
Void space of the aquifer = $V\phi$ = (1 m^3)(35%)= 0.35 m^3 = 350 L.
Total volume of water that is equivalent to 100 pore volumes = (100)(350) = 35,000 L.
The minimum required nutrient concentration = 184 g ÷ 35,000 L = 5.3×10^{-3} g/L ~ 0.0005 % by weight.

Discussion. The concentration 0.0005% by weight is the theoretical amount. In real applications, one may add more to compensate the loss due to adsorption to the aquifer material before reaching the plume. This will make the nutrient concentration fall in the typical range of 0.005 to 0.02% by weight.

VI.3.2 Air sparging

Air sparging is an emerging in situ remediation technology that involves the injection of air (sometimes oxygen) into the saturated zone. The injected air travels through the aquifer, moves upward through the capillary fringe and the vadose zone, and is then collected by the vadose zone soil venting network. The injected air (or oxygen) serves the following functions: (1) volatilizes the dissolved VOCs in the groundwater, (2) supplies oxygen to the aquifer for bioremediation, (3) volatilizes the VOCs in the capillary zone as it moves upward, and (4) volatilizes the VOCs in the vadose zone.

Amount of oxygen added to the groundwater

As illustrated in the previous sections, the amount of oxygen carried into the contaminant plume by the reinjected water, which has been saturated with air or pure oxygen, cannot meet the oxygen demand for in situ bioremediation. An air sparging process continuously brings air (or oxygen) directly into the plume. Consequently, supplying oxygen is one of the main functions of air sparging. Oxygen transfer efficiency (E) is often used to evaluate the efficacy of aeration, and it is defined as

$$\text{Oxygen Transfer Efficiency } (E) = \frac{\text{Rate of Oxygen Dissolution}}{\text{Rate of Oxygen Applied}} \qquad \text{[Eq. VI.3.2]}$$

Many studies have been conducted on aeration for water and wastewater treatment, but little information is available regarding the air sparging of the groundwater aquifer. Nevertheless, the oxygen transfer efficiency of air sparging should be much lower than that of the well-controlled aeration process in water or wastewater, in which the transfer efficiency is normally at a few percent or less.

Example VI.3.2A Determine the rate of oxygen addition by air sparging

Three air sparging wells were installed into the contaminant plume of the aquifer described in Example V.3.1A. The injection air flow rate to each well is 5 ft^3/min. Assuming the oxygen transfer efficiency is 2%, determine the rate of oxygen addition to the aquifer through each sparging well. What would be the equivalent injection rate of water that is saturated with air?

Solution:

a. The oxygen concentration in the ambient air is approximately 21% by volume, which is equal to 210,000 ppmV. Eq. II.1.1 or II.1.2 can be used to convert it to a mass concentration:

$$1\,\text{ppmV} = \frac{\text{MW}}{385} \times 10^{-6}\,[\text{lb/ft}^3] \quad \text{at}\,68°\text{F}$$

$$= \frac{32}{385} \times 10^{-6} = 0.083 \times 10^{-6}\ \text{lb/ft}^3 \qquad \text{[Eq. II.1.2]}$$

Therefore,

$$210{,}000\ \text{ppmV} = (210{,}000)(0.083 \times 10^{-6}) = 0.0175\ \text{lb/ft}^3$$

b. The rate of oxygen injected in each well = $(G)(Q)$ = (0.0175 lb/ft^3)(5 ft^3/min) = 0.0875 lb/min = 126 lb/day.
The rate of oxygen dissolved into the plume through air injection in each well (using Eq. VI.3.2) = (126 lb/day)(2%) = 2.52 lb/day.
c. The DO of the air-saturated reinjection water is approximately 9 mg/L. The required water reinjection rate to supply 2.52 lb/day of oxygen can be found as 2.52 lb/day = (2.52 lb/day)(454,000 mg/lb) = QC = Q (9 mg/L). Thus, Q = 127,000 L/day = 23.3 gpm.

Discussion

1. The oxygen transfer efficiency of 2% means that only 2% of the total oxygen added into the aquifer will dissolve into the aquifer. Although the oxygen transfer is relatively low, the 98% of the injected oxygen, that is not dissolved, is still usable as the oxygen source for bioremediation in the vadose zone.
2. Despite of the low oxygen transfer efficiency, the air sparging still adds a significant amount of oxygen to the aquifer. With regard to oxygen addition, an air injection rate of 5 ft^3/min at an oxygen transfer efficiency of 2% is equivalent to reinjection of air-saturated water at 23.3 gpm.

Air injection pressure

Air injection pressure is an important component for design of the air sparging process. The applied air injection pressure should overcome at least (1) the hydrostatic pressure corresponding to the water column height above the injection point and (2) the "air entry pressure," which is equivalent to the capillary pressure necessary to induce air into the saturated media.

$$P_{injection} = P_{hydrostatic} + P_{capillary} \qquad \text{[Eq. VI.3.3]}$$

Reported values of injection pressures range from 1 to 8 psig.[3]

The following procedure can be used to determine the minimum air injection pressure:

Step 1: Determine the water column height above the injection point. Convert the water column height to pressure units by using the following formula:

$$P_{hydrostatic} = \rho g h_{hydrostatic} \quad \text{[Eq. VI.3.4]}$$

where ρ is the mass density of water and g is the gravitational constant.

Step 2: Use Table II.1.B to estimate the pore radius of the aquifer media and then use Eq. II.1.9 to determine height of capillary rise (or obtain the capillary height from Table II.1.B directly). Convert the capillary height to the capillary pressure by using the following formula:

$$P_{capillary} = \rho g h_{capillary} \quad \text{[Eq. VI.3.5]}$$

Step 3: The minimum air injection pressure is the sum of the above two pressure components.

Information needed for this calculation

- Depth of the injection point, $h_{hydrostatic}$
- Mass density of water, ρ
- Geology of the aquifer material or the pore size of the matrix

---

Example VI.3.2B Determine the required air injection pressure of air sparging

Three air sparging wells were installed into the contaminant plume of the aquifer described in Example VI.3.1A. The injection air flow rate to each well is 5 ft³/min. The height of the water column above the air injection point is 10 ft. The aquifer matrix consists mainly of coarse sand. Determine the minimum air injection pressure required. Also, for the purpose of comparison, determine the air injection pressure if the aquifer formation is clayey.

Solution:

a. Use Eq. VI.3.4 to convert the water column height to pressure units as

$$P_{hydrostatic} = \rho g h_{hydrostatic} = \left(62.4\frac{\text{lb}_\text{m}}{\text{ft}^3}\right)\left(32.2\frac{\text{ft}}{\text{s}^2}\right)(10\,\text{ft})\left[\frac{\text{lb}_\text{f}}{32.2\,\text{lb}_\text{m} - \text{ft/s}^2}\right]$$

$$= 624\frac{\text{lb}_\text{f}}{\text{ft}^2} = 4.33\frac{\text{lb}_\text{f}}{\text{in}^2} = 4.33\,\text{psi}$$

Note that (1) the density of water at 60°F is 62.4 lb_m/ft^3. In other words, the specific weight of water is 62.4 lb_f/ft^3. (2) The water column height of 33.9 ft at 60°F is equivalent to one atmospheric pressure or 14.7 psi.

b. From Table II.1.B, pore radius of fine sand media is 0.05 cm. Use Eq. II.1.9 to determine the height of capillary rise:

$$h_c = \frac{0.153}{r} = \frac{0.153}{0.05} = 3.06 \text{ cm} = 0.1 \text{ ft}$$

Use the discussions in (a) to convert the capillary rise to the capillary pressure:

$$P_{capillary} = \left(\frac{0.1\,\text{ft}}{33.9\,\text{ft}}\right)(14.7\,\text{psi}) = 0.04\,\text{psi}$$

c. Use Eq. VI.3.3 to determine the minimum air injection pressure:

$$P_{injection} = P_{hydrostatic} + P_{capillary} = 4.33 + 0.04 = 4.37 \text{ psig}$$

d. If the aquifer formation is clayey, then the pore radius is 0.0005 cm from Table II.1.B. Use Eq. II.1.9 to determine the height of the capillary rise:

$$h_c = \frac{0.153}{r} = \frac{0.153}{0.0005} = 306\,\text{cm} = 10\,\text{ft}$$

Use the discussions in (a) to convert the capillary rise to the capillary pressure:

$$P_{capillary} = \left(\frac{10\,\text{ft}}{33.9\,\text{ft}}\right)(14.7\,\text{psi}) = 4.33\,\text{psi}$$

Use Eq. VI.3.3 to determine the minimum air injection pressure:

$$P_{injection} = P_{hydrostatic} + P_{capillary} = 4.33 + 4.33 = 8.66 \text{ psig}$$

Discussion

1. The actual air injection pressure should be larger than the minimum air injection pressure calculated above to cover the system pressure loss such as head loss in the pipeline, fittings, and injection (or diffuser) head.

2. For sandy aquifers the air entry pressure is negligible compared to the hydrostatic pressure. However, for clayey aquifers the entry pressure is of the same order of magnitude as the hydrostatic pressure.
3. The calculated injection pressures are in the ball park of the reported field values, 1 to 8 psig.

Power requirement for air injection

Theoretical horsepower requirements ($hp_{theoretical}$) of gas compressors for an ideal gas undergoing an isothermal compression (PV = constant) can be expressed as[5]

$$hp_{theoretical} = 3.03 \times 10^{-5} P_1 Q_1 \ln \frac{P_2}{P_1} \qquad \text{[Eq.VI.3.6]}$$

where P_1 = intake pressure, lb_f/ft^2, P_2 = final delivery pressure, lb_f/ft^2, and Q_1 = air flow rate at the intake condition, ft^3/min.

For an ideal gas undergoing an isentropic compression (PV^k = constant), the following equation applies for single-stage compressor[5]

$$hp_{theoretical} = \frac{3.03 \times 10^{-5}\, k}{k-1} P_1 Q_1 \left[\left(\frac{P_2}{P_1} \right)^{(k-1)/k} - 1 \right] \qquad \text{[Eq.VI.3.7]}$$

where k is the ratio of specific heat of gas at constant pressure to specific heat of gas at constant volume. For air sparging applications, it is appropriate to use $k = 1.4$.

For reciprocating compressors, the efficiencies (E) are generally in the range of 70 to 90% for isentropic and 50 to 70% for isothermal compression. The actual horsepower requirement can be found as

$$hp_{actual} = \frac{hp_{theoretical}}{E} \qquad \text{[Eq.VI.3.8]}$$

Example VI.3.2C Determine the required air injection pressure of air sparging

Three air sparging wells were installed into the contaminant plume of the aquifer described in Example VI.3.1A. The injection air flow rate to each well is 5 ft^3/min. A compressor is to serve all three wells. Head loss of the piping system and the injection head was found to be 1 psi. Using the calculated air injection pressure from Example VI.3.2B, determine the required horsepower of the compressor.

Solution:

a. The required injection pressure = the final delivery of the compressor, P_2 = minimum injection pressure + head loss = 4.37 + 1.0 = 5.37 psig = (5.37 + 14.7) psia = 20.1 psia = (20.1)(144) = 2890 lb_f/ft^2.

b. Assuming isothermal expansion, use Eq. VI.3.6 to determine the theoretical power requirement as

$$hp_{theoretical} = 3.03 \times 10^{-5} P_1 Q_1 \ln \frac{P_2}{P_1}$$

$$= 3.03 \times 10^{-5} [(14.7)(144)][(3)(5)] \ln \frac{2890}{(14.7)(144)} = 0.3hp$$

Assuming an isothermal efficiency of 60%, the actual horsepower required is determined by using Eq. VI.3.8:

$$hp_{actual} = \frac{hp_{theoretical}}{E} = \frac{0.3}{60\%} = 0.5\,hp$$

c. Assuming isothermal expansion, use Eq. VI.3.7 to determine the theoretical power requirement as

$$hp_{theoretical} = \frac{3.03 \times 10^{-5} k}{k-1} P_1 Q_1 \left[\left(\frac{P_2}{P_1} \right)^{(k-1)/k} - 1 \right]$$

$$= \frac{(3.03 \times 10^{-5})(1.4)}{1.4-1} [(14.7)(144)][(5)(3)] \left[\left(\frac{2890}{(14.7)(144)} \right)^{(1.4-1)/1.4} - 1 \right] = 0.31\,hp$$

Assuming an isentropic efficiency of 80%, the actual horsepower required is determined by using Eq. VI.3.8 as

$$hp_{actual} = \frac{hp_{theoretical}}{E} = \frac{0.31}{80\%} = 0.4\,hp$$

Discussion. The energy necessary for an isentropic compression is generally greater than that for an equivalent isothermal compression. However, the difference between the inlet and final discharge pressures in most air sparging applications is relatively small. Consequently, the theoretical power requirements for the isothermal and isentropic compressions should be very similar, as illustrated in this example.

References

1. U.S. EPA, Site Characterization for Subsurface Remediation, EPA/625/R-91/026, Office of Research and Development, U.S. EPA, Washington, DC, 1991.
2. Javandel, I. and Tsang, C.-F., Capture–zone type curves: a tool for aquifer cleanup, *Groundwater,* 24(5), 616–625, 1986.
3. Johnson, R. L., Johnson, P. C., McWhorter, D. B., Hinchee, R. E., and Goodman, I., An overview of in situ air sparging, *Ground Water Monitor. Rev.,* Fall, 127–135, 1993.
4. Metcalf & Eddy, Inc., *Wastewater Engineering,* 3rd ed., McGraw–Hill, New York, 1991.
5. Peters, M. S. and Timmerhaus, K. D., *Plant Design and Economics for Chemical Engineers,* 4th ed., McGraw–Hill, New York, 1991.

chapter seven

VOC-laden air treatment

Remediation of contaminated soil and groundwater often results in transferring organic contaminants into the air phase. Development and implementation of an air emission control strategy should be an integral part of the overall remediation program. Air emission control may affect the cost-effectiveness of a specific remedial alternative.

Common sources of VOC-laden off-gas from soil/groundwater remediation activities include soil vapor extraction, air sparging, air stripping, solidification/stabilization, and bioremediation. This chapter illustrates the design calculations for commonly used treatment technologies: activated carbon adsorption, direct incineration, catalytic incineration, IC engines, and biofiltration.

VII.1 Activated carbon adsorption

Process description

Activated carbon adsorption is one of the most commonly used air pollution control processes for reducing VOC emission from soil/groundwater remediation. The process is very effective in removing a wide range of VOCs. The most common form of activated carbon for this type of application is granular activated carbon (GAC).

Activated carbon has a fixed capacity or a limited number of active adsorption sites. Once the adsorbing contaminants occupy most of the available sites, the adsorption efficiency will drop significantly. If the operation is continued beyond this point, the breakthrough point will be reached and the effluent concentration will increase sharply. Eventually, carbon would be "saturated," "exhausted," or "spent" when all sites are occupied. The spent carbon needs to be regenerated or disposed of.

Two pretreatment processes are often required to optimize the performance of GAC systems. The first is cooling, and the other is dehumidification. Adsorption of VOCs is generally exothermic, which is favored by lower temperatures. As a rule of thumb, the waste air stream needs to be cooled down below 130°F. Water vapor will compete with VOCs in the waste air stream for available adsorption sites. The relative humidity of the waste air stream generally should be reduced to 50% or less.

GAC sizing criteria

Various GAC adsorber designs are commercially available. Two of the most common ones are (1) canister systems with off-site regeneration and (2) multiple-bed systems with on-site batch regeneration (while some of the adsorbers are in adsorption cycle, the others are in regeneration cycle).

Sizing of the GAC systems depends primarily on the following parameters:

1. Volumetric flow rate of VOC-laden gas stream
2. Concentration or mass loading of VOCs
3. Adsorption capacity of GAC
4. Desired GAC regeneration frequency

The flow rate determines the size or cross-sectional area of the GAC bed, the size of the fan and motor, and the duct diameter. The other three, mass loading, GAC adsorption capacity, and regeneration frequency, determine the amount of GAC required for a specific project. Design of vapor-phase activated carbon systems is basically the same as that for liquid-phase activated carbon systems, as described in Section VI.2.

VII.1.1 Adsorption isotherm and adsorption capacity

The adsorption capacity of GAC depends on the type of GAC and the type of VOC compounds and their concentration, temperature, and presence of other species competing for adsorption. At a given temperature, a relationship exists between the mass of the VOC adsorbed per unit mass GAC and the concentration (or partial pressure) of VOC in the waste air stream. For most of the VOCs, the adsorption isotherms can be fitted well by a power curve, also known as the Freundlich isotherms (also see Eq. VI.2.2):

$$q = a(P_{VOC})^m \quad \text{[Eq. VII.1.1]}$$

where q = equilibrium adsorption capacity, lb VOC/lb GAC, P_{VOC} = partial pressure of VOC in the waste air stream, psi, and a, m = empirical constants.

The empirical constants of the Freundlich Isotherms for selected VOCs are listed in Table VII.1.A. It should be noted that the values of these empirical constants are for a specific type of GAC only and should not be used outside the specified range.

The actual adsorption capacity in the field applications should be lower than the equilibrium adsorption capacity. Normally, design engineers take

Table VII.1.A Empirical Constants for Selected Adsorption Isotherms

Compounds	Adsorption Temperature (°F)	a	m	Range of P_{VOC} (psi)
Benzene	77	0.597	0.176	0.0001–0.05
Toluene	77	0.551	0.110	0.0001–0.05
m-Xylene	77	0.708	0.113	0.0001–0.001
	77	0.527	0.0703	0.001–0.05
Phenol	104	0.855	0.153	0.0001–0.03
Chlorobenzene	77	1.05	0.188	0.0001–0.01
Cyclohexane	100	0.508	0.210	0.0001–0.05
Dichloroethane	77	0.976	0.281	0.0001–0.04
Trichloroethane	77	1.06	0.161	0.0001–0.04
Vinyl chloride	100	0.20	0.477	0.0001–0.05
Acrylonitrile	100	0.935	0.424	0.0001–0.05
Acetone	100	0.412	0.389	0.0001–0.05

From U.S. EPA, Control Technologies for Hazardous Air Pollutants, EPA/625/6-91/014, U.S. EPA, Washington, DC, 1991.

25 to 50% of the equilibrium value as the design adsorption capacity as a factor of safety. Therefore,

$$q_{actual} = (50\%)(q_{theoretical}) \qquad \text{[Eq. VII.1.2]}$$

The maximum amount of contaminants that can be removed or held ($M_{removal}$) by a given amount of GAC can be determined as

$$\begin{aligned} M_{removal} &= (q_{actual})(M_{GAC}) \\ &= (q_{actual})[(V_{GAC})(\rho_b)] \end{aligned} \qquad \text{[Eq. VII.1.3]}$$

where M_{GAC} is the mass, V_{GAC} is the volume, and ρ_b is the bulk density of the GAC, respectively.

The following procedure can be used to determine the adsorption capacity of a GAC adsorber:

Step 1: Determine the theoretical adsorption capacity by using Eq. VII.1.1.
Step 2: Determine the actual adsorption capacity by using Eq. VII.1.2.
Step 3: Determine the amount of activated carbon in the adsorber.
Step 4: Determine the maximum amount of contaminants that can be held by the adsorber using Eq. VII.1.3.

Information needed for this calculation

- Adsorption isotherm
- Contaminant concentration of the influent waste air stream, P_{VOC}

- Volume of the GAC, V_{GAC}
- Bulk density of the GAC, ρ_b

Example VII.1.1 *Determine the capacity of a GAC adsorber*

The off-gas from a soil venting project is to be treated by GAC adsorbers. The *m*-xylene concentration in the off-gas is 800 ppmV. The air flow rate out of the extraction blower is 200 cfm, and the temperature of the air is ambient. Two 1000-lb activated carbon adsorbers are proposed. Determine the maximum amount of *m*-xylene that can be held by each GAC adsorber before regeneration. Use the isotherm data in Table VII.1.A.

Solution:

a. Convert the xylene concentration from ppmV to psi as

$$P_{VOC} = 800 \text{ ppmV} = 800 \times 10^{-6} \text{ atm} = 8.0 \times 10^{-4} \text{ atm}$$
$$= (8.0 \times 10^{-4} \text{ atm})(14.7 \text{ psi/atm}) = 0.0118 \text{ psi}$$

Obtain the empirical constants for the adsorption isotherm from Table VII.1.A and then apply Eq. VII.1.1 to determine the equilibrium adsorption capacity as

$$q = a(P_{VOC})^m = (0.527)(0.0118)^{0.0703} = 0.386 \text{ lb/lb}$$

b. The actual adsorption capacity can be found by using Eq. VII.1.2 as

$$q_{actual} = (50\%)q_{theoretical} = (50\%)(0.386) = 0.193 \text{ lb/lb}$$

c. Amount of xylene that can be retained by an adsorber before the GAC becomes exhausted = (amount of the GAC)(actual adsorption capacity) = (1000 lbs/unit)(0.193 lb xylene/lb GAC) = 193 lb xylene/unit.

Discussion

1. The adsorption capacity of vapor-phase GAC is typically in the neighborhood of 0.1 lb/lb (or 0.1 kg/kg), which is much higher than the adsorption capacity of liquid-phase GAC, typically in the neighborhood of 0.01 lb/lb.
2. Care should be taken to use matching units for P_{VOC} and q in the isotherm equations.
3. The influent contaminant concentration in the air stream, not the effluent concentration, should be used in the isotherm equations to determine the adsorption capacity.
4. There are two sets of empirical constants for *m*-xylene; one should always check the applicable range for the empirical constants.

VII.1.2 Cross-sectional area and height of GAC adsorbers

To achieve efficient adsorption, the air flow rate through the activated carbon should be kept as low as possible. The practical design air flow velocity is often selected to be 60 ft/min or less, and 100 ft/min is considered as the maximum value. This design parameter is often used to determine the required cross-sectional area of the GAC adsorbers (A_{GAC}):

$$A_{GAC} = \frac{Q}{Air\ Flow\ Velocity} \qquad \text{[Eq. VII.1.4]}$$

where Q is the air flow velocity. The design height of the adsorber is normally 2 ft or greater to provide a sufficiently large adsorption zone.

Example VII.1.2 Required cross-sectional area of GAC adsorbers

Referring to the remediation project described in Example VII.1.1, the 1000-lb GAC units are out of stock. To avoid delay of remediation, off-the-shelf 55-gal activated carbon units are proposed on an interim basis. The type of carbon in the 55-gal units is the same as that in the 1000-lb units. The vendor also provided the following information regarding the units:

Diameter of carbon packing bed in each 55-gal drum = 1.5 ft
Height of carbon packing bed in each 55-gal drum = 3 ft
Bulk density of the activated carbon = 28 lb/ft^3

Determine (a) the amount of activated carbon in each 55-gal unit, (b) the amount of xylene that each unit can remove before being exhausted, and (c) the minimum number of the 55-gal units needed.

Solution:

a. Volume of the activated carbon inside a 55-gal drum = $(\pi r^2)(h)$

$$= (\pi)[(1.5/2)^2](3) = 5.3 \text{ ft}^3$$

Amount of the activated carbon inside a 55-gal drum = $(V)(\rho_b)$

$$= (5.3 \text{ ft}^3)(28 \text{ lb/ft}^3) = 148 \text{ lbs}$$

b. Amount of xylene that can be retained by a drum before the GAC becomes exhausted

$$= \text{(amount of the GAC)(actual adsorption capacity)}$$
$$= (148 \text{ lbs/drum})(0.193 \text{ lb xylene/lb GAC}) = 28.6 \text{ lb xylene/drum}$$

c. Assuming a design air flow velocity of 60 ft/min, the required cross-sectional area for the GAC adsorption can be found by using Eq. VII.1.4 as

$$A_{GAC} = \frac{Q}{Air\ Flow\ Velocity} = \frac{200}{60} = 3.33\,\text{ft}^2$$

If the adsorption system is tailor made, then a system with a cross-sectional area of 3.33 ft^2 will do the job. However, the off-the-shelf 55-gal drums are to be used, so we need to determine the number of drums that will provide the required cross-sectional area.
Area of the activated carbon inside a 55-gal drum = $(\pi r^2) = (\pi)[(1.5/2)^2]$ = 1.77 ft^2/drum.
Number of drums in-parallel to meet the required hydraulic loading rate = (3.33 ft^2) ÷ (1.77 ft^2/drum) = 1.88 drums.
So, use two drums in parallel to provide the required cross-sectional area. The total cross-sectional area of two drums is equal to 3.54 ft^2 (= 1.77 × 2).

Discussion

1. The bulk density of vapor-phase GAC is typically in the neighborhood of 30 lb/ft^3. The amount of activated carbon in a 55-gal drum is approximately 150 pounds.
2. The minimum number of 55-gal drums for this project is two to meet the air flow velocity requirement. The actual number of drums should be more to meet the monitoring requirements or the desirable frequency of change-out. If multiple GAC adsorbers are used, the adsorbers are often arranged in series and/or in parallel. If two adsorbers are arranged in series, the monitoring point can be located at the effluent of the first adsorber. A high effluent concentration from the first adsorber indicates that this adsorber is reaching its capacity. The first adsorber is then taken off-line, and the second adsorber is shifted to be the first adsorber. Consequently, the capacity of both adsorbers can be fully utilized and the compliance requirements can also be met. If there are two parallel streams of adsorbers, one stream can always be taken off-line for regeneration or maintenance, and the continuous operation of the system is secured.

VII.1.3 Contaminant removal rate by the activated carbon adsorber

The removal rate by a GAC adsorber ($R_{removal}$) can be calculated by using the following formula:

$$R_{removal} = (G_{in} - G_{out})Q \qquad \text{[Eq. VII.1.5]}$$

In practical applications, the effluent concentration (G_{out}) is kept below the discharge limit, which is often very low. Therefore, for a factor of safety, the term of G_{out} can be deleted from Eq. VII.1.5 in design. The mass removal rate is then the same as the mass loading rate ($R_{loading}$):

$$R_{removal} \sim R_{loading} = (G_{in})Q \qquad \text{[Eq. VII.1.6]}$$

The mass loading rate is nothing but the multiplication product of the air flow rate and the contaminant concentration. As mentioned earlier, the contaminant concentration in the air is often expressed in ppmV or ppbV. In the mass loading rate calculation, the concentration has to be converted to mass concentration units as

$$\begin{aligned} 1\,\text{ppmV} &= \frac{\text{MW}}{22.4}[\text{mg/m}^3] \quad \text{at } 0°\text{C} \\ &= \frac{\text{MW}}{24.05}[\text{mg/m}^3] \quad \text{at } 20°\text{C} \\ &= \frac{\text{MW}}{24.5}[\text{mg/m}^3] \quad \text{at } 25°\text{C} \end{aligned} \qquad \text{[Eq. VII.1.7]}$$

or

$$\begin{aligned} 1\,\text{ppmV} &= \frac{\text{MW}}{359} \times 10^{-6}\,[\text{lb/ft}^3] \quad \text{at } 32°\text{C} \\ &= \frac{\text{MW}}{385} \times 10^{-6}\,[\text{lb/ft}^3] \quad \text{at } 68°\text{C} \\ &= \frac{\text{MW}}{392} \times 10^{-6}\,[\text{lb/ft}^3] \quad \text{at } 77°\text{C} \end{aligned} \qquad \text{[Eq. VII.1.8]}$$

where MW is the molecular weight of the compound.

Example VII.1.3 *Determine the mass removal rate by the GAC adsorbers*

Referring to the remediation project described in Example VII.1.2, the discharge limit for xylene is 100 ppbV. Determine the mass removal rate by the two 55-gal GAC units.

Solution:

a. Use Eq. VII.1.8 to convert the ppmV concentration to lb/ft^3. Molecular weight of xylene ($C_6H_4(CH_3)_2$) = 12 × 8 + 1 × 10 = 106.

$$1\,\text{ppmV} = \frac{106}{392} \times 10^{-6} = 0.27 \times 10^{-6}\ \text{lb/ft}^3 \quad \text{at}\ 77°\text{F}$$

800 ppmV = (800)(0.27 × 10^{-6}) = 2.16 × 10^{-4} lb/ft^3.

b. Use Eq. VII.1.6 to determine the mass removal rate:

$$R_{removal} \sim (G_{in})Q =$$
$$(2.16 \times 10^{-4}\ \text{lb/ft}^3)(200\ \text{ft}^3/\text{min}) = 0.65\ \text{lb/min} = 93\ \text{lb/d}$$

VII.1.4 Change-out (or regeneration) frequency

Once the activated carbon reaches its capacity, it should be regenerated or disposed of. The time interval between two regenerations or the expected service life of a fresh batch of activated carbon can be found by dividing the capacity of activated carbon with the contaminant removal rate ($R_{removal}$) as

$$T = \frac{M_{removal}}{R_{removal}} \qquad \text{[Eq. VII.1.9]}$$

Example VII.1.4 Determine the change-out (or regeneration) frequency of the GAC adsorbers

Referring to the remediation project described in Example VII.1.3, the discharge limit for xylene is 100 ppbV. Determine the service life of the two 55-gal GAC units.

Solution:

As shown in Example VII.1.2, the amount of xylene that each drum can retain before being exhausted is 28.6 lbs. Use Eq. VII.1.9 to determine the service life of two drums:

$$T = \frac{M_{removal}}{R_{removal}} = \frac{(2)(28.6\,\text{lb})}{0.65\,\text{lb/min}} = 88\,\text{min} < 1.5\,\text{hrs}$$

Discussion

1. Although two drums in parallel can provide a sufficient cross-sectional area for adequate air flow velocity, the relatively high contaminant concentration makes the service life of the two 55-gal drums unacceptably short.
2. A 55-gal activated carbon drum normally costs several hundred dollars. In this example, two drums last less than 90 minutes. The labor and disposal costs should also be added, and it makes this option prohibitive. A GAC system with on-site regeneration or other treatment alternatives should be considered.

VII.1.5 Amount of carbon required (on-site regeneration)

If the concentration of the waste air stream is high, a GAC system with on-site regeneration capability would become an attractive option. The amount of GAC required for on-site regeneration depends on the mass loading, the adsorption capacity of GAC, the adsorption time between two regenerations, and the ratio between the number of GAC beds in regeneration cycle and the number of GAC beds in adsorption cycle. It can be determined by using the following formula:

$$M_{GAC} = \frac{R_{removal} T_{ad}}{q}\left[1 + \frac{N_{des}}{N_{ad}}\right] \qquad \text{[Eq. VII.1.10]}$$

where M_{GAC} = total amount of GAC required, T_{ad} = adsorption time between two regeneration (desorption), N_{ads} = number of GAC beds in adsorption phase, and N_{des} = number of GAC beds in desorption (regeneration) phase.

Example VII.1.5 Determine the amount of GAC required for on-site regeneration

Referring to the remediation project described in Example VII.1.3, an on-site regeneration GAC is proposed to deal with the high contaminant loading. The system consists of three adsorbers. Two of the three adsorbers are in adsorption cycle and the other one is in regeneration cycle. The adsorption cycle time is two hours. Determine the amount of GAC required for this system.

Solution:

The total amount of GAC required in all three adsorbers can be determined by using Eq. VII.1.10 as

$$M_{GAC} = \frac{R_{removal} T_{ad}}{q}\left[1 + \frac{N_{des}}{N_{ad}}\right]$$

$$= \frac{(0.65\,\text{lb/min})(120\,\text{min})}{(0.193\,\text{lb/lb})}\left[1 + \frac{1}{2}\right] = 606\,\text{lbs}$$

So, 202 pounds of GAC in each bed are required.

VII.2 Thermal oxidation

Thermal processes are commonly used to treat VOC-laden air. Thermal oxidation, catalytic oxidation, and internal combustion (IC) engines are popular thermal processes. The key components of thermal treatment system design are referred to as the "three T's," which are combustion temperature, residence time (also called "retention time" or "dwell time"), and turbulence. They basically determine a reactor's size and its destruction efficiency. For example, to achieve good thermal destruction, the VOC-laden gas should be held in a thermal oxidizer for a sufficient residence time (normally 0.3 to 1.0 seconds) at a temperature at least 100°F above the autoignition temperatures of the compounds in the VOC-laden gas stream. In addition, sufficient turbulence must be maintained in the oxidizer to assure good mixing and complete combustion of the contaminants. Other important parameters to be considered include influent concentration (heating value) and auxiliary fuel and supplementary air requirements.

Discussions on the combustion basics for thermal oxidation will be presented here and they are applicable to other thermal processes with little modification needed.

VII.2.1 Air flow rate vs. temperature

The volumetric air flow rate is commonly expressed in ft^3/min, i.e., cubic feet per minute (cfm). Since the volumetric flow rate of an air stream is a function of temperature and the air stream undergoes zones of different temperatures in a thermal process, the air flow rate is further shown as actual cfm (acfm) or standard cfm (scfm). The unit of acfm refers to the volumetric flow rate under the actual temperature, while scfm is the flow rate at standard conditions. The standard conditions are the basis for comparison. Unfortunately, the definition of the standard conditions is not universal. For U.S. EPA the standard conditions are at 77°F (25°C) and 1 atmospheric pressure; however, it is 68°F (20°C) and 1 atm for the South Coast Air Quality Management Districts in southern California. In addition, 60°F is also commonly used in the literature or in books as the temperature for the standard conditions. One should follow the regulatory requirements and use the appropriate reference temperature for a specific project. A standard temperature of 77°F is used in this chapter, unless otherwise specified.

Conversions between acfm and scfm for a given air stream can be easily made using the following formula which assumes that the ideal gas law is valid:

$$\frac{Q_{actual@temperature\,T,\,in\,acfm}}{Q_{standard,\,in\,scfm}} = \frac{460+T}{460+77} \qquad \text{[Eq. VII.2.1]}$$

where T is the actual temperature in °F and the addition of 460 is to convert the temperature from °F to degree Rankine.

Example VII.2.1 Conversion between the actual and standard air flow rates

A thermal oxidizer was used to treat the off-gas from a soil venting process. To achieve the required removal efficiency the oxidizer was operated at

1400°F. The flow rate at the exit of the oxidizer was 550 ft³/min. What would be the exit flow rate expressed in scfm? The temperature of the effluent air from the final discharge stack was 200°F. If the diameter of the final stack was 4 in, determine the air flow velocity from the discharge stack.

Solution:

a. Use Eq. VII.2.1 to convert acfm to scfm as

$$\frac{Q_{actual@temperature T, in acfm}}{Q_{standard, in scfm}} = \frac{460+T}{460+77} = \frac{460+1400}{460+77} = \frac{550}{Q_{standard, in scfm}}$$

So,

$$Q = 158.8 \text{ scfm.}$$

b. Use Eq. VII.2.1 to determine the flow rate from the stack:

$$\frac{Q_{actual@temperature T, in acfm}}{Q_{standard, in scfm}} = \frac{460+T}{460+77} = \frac{460+200}{460+77} = \frac{Q_{actual@temperature T, in acfm}}{158.8}$$

So, Q = 195.2 acfm @ 200°F.
The discharge velocity, $v = Q/A = Q \div (\pi r^2)$ = 195.2 ft³/min ÷ [$\pi(2/12)^2$ ft²] = 2240 ft/min.

Discussion. If the actual flow rate at one temperature is known, it can be used to determine the flow rate at another temperature by using the following formula:

$$\frac{Q_{actual} @ T_1}{Q_{actual} @ T_2} = \frac{460+T_1}{460+T_2} \qquad \text{[Eq. VII.2.2]}$$

The stack flow rate in this example can be directly determined from the exit flow rate from the oxidizer as

$$\frac{Q_{actual} @ T_1}{Q_{actual} @ T_2} = \frac{460+T_1}{460+T_2} = \frac{550}{Q_{actual} @ 200°\text{F}} = \frac{460+1400}{460+200}$$

Thus, Q_{actual} @ 200°F = 195.2 acfm.

VII.2.2 *Heating values of an air stream*

Organic compounds generally contain high heating values. These organic compounds can also serve as energy sources for combustion. The higher the organic concentration in a waste stream, the higher the heat content is and the lower the requirement for auxiliary fuel would be. If the heating value of a compound is not available, the following Dulong's formula can be used:

$$\text{Heating value (in Btu/lb)} = 145.4\,C + 620\left(H - \frac{O}{8}\right) + 41S \qquad \text{[Eq. VII.2.3]}$$

where *C*, *H*, *O*, and *S* are the percentages by weight of these elements in the compound. Eq. VII.2.3 can also be used to estimate the heating value of a solid waste. The heating value of an air stream containing organics can be determined by

Heating value of an air stream containing VOCs (in Btu/scf) = VOCs heating value (in Btu/lb) × mass concentration of the VOC (lb/scf) [Eq. VII.2.4]

We can divide the heating value of a waste air stream in Btu/scf by the density of the air to obtain the heating value in Btu/lb.

Heating value of an air stream containing VOCs (in Btu/lb) = heating value (in Btu/scf) ÷ density of the air stream (lb/scf) [Eq. VII.2.5]

The density of an air stream under standard conditions can be found as

$$\text{Density of an air stream (in lb/scf)} = \frac{\text{Molecular Weight}}{392} \qquad \text{[Eq. VII.2.6]}$$

Since the air consists mainly of 21% oxygen (molecular weight = 32) and 79% nitrogen (molecular weight = 28), people normally use 29 as the molecular weight of the air. Consequently, the density of the air is 0.0739 lb/scf (= 29/392). This value can also be used for VOC-laden air, provided the VOC concentrations are not extremely high.

Example VII.2.2 Estimate the heating value of an air stream

Referring to the remediation project described in Example VII.1.3, a thermal oxidizer is also considered to treat the off-gas. Estimate the heating value of the air stream that contains 800 ppmV of xylene.

Solution:

a. Use Dulong's formula (Eq. VII.2.3) to estimate the heating value of pure xylene.
Molecular weight of xylene ($C_6H_4(CH_3)_2$) = 12 × 8 + 1 × 10 = 106.
Weight percentage of C = (12 × 8) ÷ 106 = 90.57%.

Weight percentage of H = (1 × 10) ÷ 106 = 9.43%.

$$\text{Heating value (in Btu/lb)} = 145.4C + 620\left(H - \frac{0}{8}\right) + 41S = 145.4(90.57)$$

$$+\ 620\left(9.43 - \frac{0}{8}\right) + 41(0) = 19{,}015.$$

b. To determine the heat content of the air containing 800 ppmV xylene, we have to determine the mass concentration of xylene in the air first (which has been previously determined in Example VII.1.3):

$$\begin{aligned} 800 \text{ ppmV of xylene} &= (800)(0.27 \times 10^{-6}) \\ &= 2.16 \times 10^{-4} \text{ lb of xylene/ft}^3 \text{ of air.} \end{aligned}$$

Use Eq. VII.2.4 to determine the heating value of the off-gas:

$$\begin{aligned} \text{Heating value (in Btu/scf)} &= 19{,}015 \text{ Btu/lb} \times (2.16 \times 10^{-4} \text{ lb/scf}) \\ &= 4.11 \text{ Btu/scf.} \end{aligned}$$

c. Use Eq. VII.2.5 to convert the heating value into Btu/lb: Heating value of an air stream containing VOCs (in Btu/lb) = 4.11 Btu/scf ÷ 0.0739 lb/scf = 55.6 Btu/lb.

Discussion

1. The heating value of xylene calculated from the Dulong's formula, 19,015 Btu/lb, is essentially the same as that listed in the literature, 18,650 Btu/lb.
2. The weight percentage of C is 90.57%, and a value of 90.57, not 0.9057, should be used in the Dulong's formula.

VII.2.3 Dilution air

Some waste air streams contain enough organic compounds to sustain burning (e.g., no auxiliary fuel is required, which means cost saving). That is why direct incineration is favorable for treating air with high organic concentrations. However, for hazardous air pollutant streams, the concentration of flammable vapors to a thermal incinerator is generally limited to 25% of the lower explosive limit (LEL), imposed by insurance companies for safety concerns. Vapor concentrations up to 40 to 50% of the LEL may be permissible, if on-line monitoring of VOC concentrations and automatic process control and shutdown are employed. Table VII.2.A lists the LELs and upper explosive limits (UELs) of some combustible compounds in air.

When the off-gas has a VOC content greater than 25% percent of its LEL (i.e., in most of the initial stages of the SVE-based cleanups), dilution air must be used to lower the contaminant concentration to below 25% of its LEL prior to incineration. The 25% LEL corresponds to a heat content of 176 Btu/lb or 13 Btu/scf in most cases.

Table VII.2.A The LEL and UEL of Some Organic Compounds in Air

Compounds	LEL, % Volume	UEL, % Volume
Methane	5.0	15.0
Ethane	3.0	3.0
Propane	2.1	9.5
n-Butane	1.8	8.4
n-Pentane	1.4	7.8
n-Hexane	1.2	7.4
n-Heptane	1.05	6.7
n-Octane	0.95	3.2
Ethylene	2.7	36
Propylene	2.4	11
1,3-Butadiene	2.0	12
Benzene	1.3	7.0
Toluene	1.2	7.1
Ethylbenzene	1.0	6.7
Xylenes	1.1	6.4
Methyl alcohol	6.7	36
Dimethyl ether	3.4	27
Acetaldehyde	4.0	36
Methyl ethyl ketone	1.9	10

From U.S. EPA, Control Technologies for Hazardous Air Pollutants, EPA/6254/6-91/014, U.S. EPA, Washington, DC, 1991.

Example VII.2.3A *Determine the heating value of an air stream at 25% of its LEL*

An off-gas contains a high level of benzene. The heating value of benzene is 18,210 Btu/lb. Determine the heating value of the off-gas corresponding to 25% of its LEL.

Solution:

a. From Table VII.2.A, the 100% LEL of benzene in air is 1.3% by volume. The 25% LEL = (25%)(1.3%) = 0.325% by volume = 3250 ppmV. Molecular weight of benzene (C_6H_6) = 12 × 6 + 1 × 6 = 78. Use Eq. VII.1.8 to convert ppmV to lb/ft³:

$$1\,\text{ppmV} = \frac{78}{392} \times 10^{-6} = 0.199 \times 10^{-6}\ \text{lb/ft}^3 \quad \text{at } 77°\text{F}$$

3250 ppmV = (3250)(0.199 × 10^{-6}) = 6.47 × 10^{-4} lb/ft³.

b. Use Eq. VII.2.4 to determine the heating value of the off-gas:

$$\text{Heating value (in Btu/scf)} = 18{,}210\ \text{Btu/lb} \times (6.47 \times 10^{-4}\ \text{lb/scf}) = 11.8\ \text{Btu/scf}$$

c. Use Eq. VII.2.5 to convert the heating value into Btu/lb:

Heating value of an air stream containing benzene (in Btu/lb)
= 11.8 Btu/scf ÷ 0.0739 lb/scf = 160 Btu/lb

Discussion. The calculated heating value, 11.8 Btu/scf or 160 Btu/lb, is very close to the general value of 13 Btu/scf or 176 Btu/lb corresponding to the 25% LEL of VOC concentration.

When dilution is required, the volumetric flow rate of the dilution air can be found as

$$Q_{dilution} = \left[\frac{H_w}{H_i} - 1\right] Q_w \qquad \text{[Eq. VII.2.7]}$$

where $Q_{dilution}$ = required dilution air, scfm, Q_w = waste air stream to be treated, scfm, H_w = heat content of the waste air stream, Btu/scf (or Btu/lb), and H_i = heat content of the desired influent entering the treatment system, Btu/scf (or Btu/lb).

Example VII.2.3B Determine the dilution air requirement

An off-gas stream (Q = 200 scfm) is to be treated by direct incineration. The heating value of the off-gas is 300 Btu/lb. The insurance policy limits the contaminant concentration in the influent air to the thermal oxidizer to 25% of its LEL. Determine the required dilution air flow rate.

Solution:

Use 176 Btu/lb as the heating value that corresponds to 25% LEL. The dilution air flow rate can be determined by using Eq. VII.2.7 as

$$Q_{dilution} = \left[\frac{H_w}{H_i} - 1\right] Q_w = \left[\frac{300}{176} - 1\right](200) = 141\,scfm$$

VII.2.4 Auxiliary air to supply oxygen

If the waste air stream has a low oxygen content (below 13 to 16%), then auxiliary air would also be used to raise the oxygen level to ensure flame stability of the burner. If the exact composition of the waste air stream is known, one can determine the stiochiometric amount of air (oxygen) for complete combustion. In general practices, excess air is added to ensure complete combustion. The following example illustrates how to determine the stiochiometric amount of air and excess air for combusting a landfill gas.

Example VII.2.4 Determine the stoichiometric air and excess air for combusting landfill gas

A landfill gas stream (60% by volume CH_4 and 40% CO_2; Q = 200 scfm) is to be treated by an incinerator. The gas is to be burned with 20% excess air at 1800°F. Determine (a) the stoichiometric amount of air required, (b) the total auxiliary air required, (c) the total influent flow rate to the incinerator, and (d) the total effluent flow rate from the incinerator.

Solution:

a. The influent flow rate of methane = (60%)(200 scfm) = 120 cfm.
The influent flow rate of carbon dioxide = (40%)(200 scfm) = 80 cfm.
The reaction for complete combustion of methane is

$$CH_4 + 2O_2 \rightarrow CO_2 + 2H_2O$$

The stoichiometric requirement of oxygen = (120 scfm)(2 moles of oxygen/one mole methane) = 240 scfm.
The stoichiometric requirement of air = (oxygen flow rate) ÷ (oxygen content in air) = (240 scfm) ÷ (21%) = 1140 scfm.

b. The total auxiliary air = (1 + 20%)(1140 scfm) = 1368 cfm.
The flow rate of nitrogen in the auxiliary air = (79%)(1370) = 1080 scfm.

c. The total influent flow rate = 120 (methane) + 80 (carbon dioxide) + 1368 (air) = 1568 scfm.

d. The flow rate of oxygen in the effluent = (20%)(240) = 48 scfm.
The flow rate of nitrogen in the effluent = the flow rate of nitrogen in the influent = 1080 scfm.
The flow rate of carbon dioxide in the effluent = carbon dioxide in the landfill gas + carbon dioxide produced from combustion = 80 + 120 (methane:carbon dioxide = 1:1) = 200 scfm.
The flow rate of water vapor in the effluent = water vapor produced from combustion (methane:water = 1:2) = (2)(120) = 240 scfm. The total effluent flow rate = 48 + 1080 + 200 + 240 = 1568 scfm.

Discussion

1. The following table summarizes the flow rate of each component in this process:

	CH_4	O_2	N_2	CO_2	H_2O
Influent (scfm)	120	2(120)(1.2) = 288	1080	80	0
Effluent (scfm)	0	288 – 240 = 48	1080	80 + 120 = 200	240

2. The flow rates of the total influent and total effluent are equal at 1568 scfm.

VII.2.5 Supplementary fuel requirements

The VOC concentration of the off-gas from soil or groundwater remediation can be very low and insufficient to support combustion. In this case auxiliary fuel is needed. The following equation can be used to determine the supplementary fuel requirement (based on natural gas):

$$Q_{sf} = \frac{D_w Q_w [C_p(1.1T_c - T_{he} - 0.1T_r) - H_w]}{D_{sf}[H_{sf} - 1.1C_p(T_c - T_r)} \qquad \text{[Eq. VII.2.8]}$$

where Q_{sf} = flow rate of supplementary fuel, scfm, D_w = density of waste air stream, lb/scf (usually 0.0739 lb/scf), D_{sf} = density of supplementary fuel, lb/scf (0.0408 lb/scf for methane), T_c = combustion temperature,°F, T_{he} = temperature of waste air stream after heat exchanger,°F, T_r = reference temperature, 77°F, C_p = mean heat capacity of air between T_c and T_r, H_w = heat content of waste air stream, Btu/lb, and H_{sf} = heating value of supplementary fuel, Btu/lb (21,600 Btu/lb for methane).

If the value of T_{he} is not specified, use the following equation to calculate T_{he}:

$$T_{he} = \left(\frac{HR}{100}\right)T_c + \left[1 - \frac{HR}{100}\right]T_w \qquad \text{[Eq. VII.2.9]}$$

where HR = heat recovery in the heat exchanger, % (If no other information is available, a value of 70% may be assumed) and T_w = temperature of the waste air stream before entering the heat exchanger,°F.

In the above equation, T_{he} is the temperature of waste air stream after heat exchanger (if no heat exchangers are employed to recuperate the heat, then $T_{he} = T_w$). The C_p value can be obtained from Figure VII.2.A.

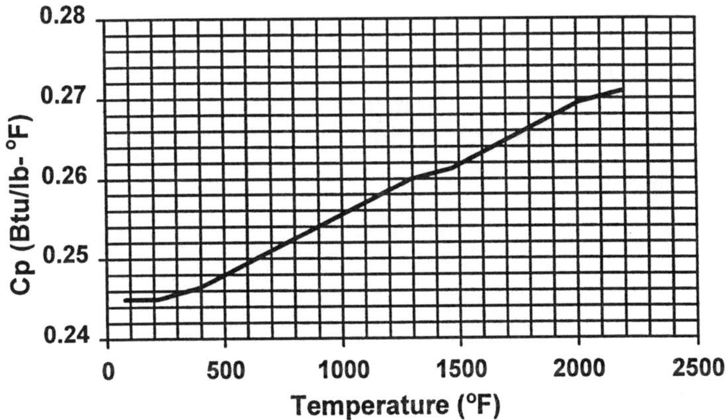

Figure VII.2.A Average specific heats of air.

Example VII.2.5 Determine the supplementary fuel requirements
Referring to the remediation project described in Example VII.2.2, an off-gas stream (Q = 200 scfm) containing 800 ppmV of xylene is to be treated by a thermal oxidizer with a recuperative heat exchanger. The combustion temperature is set at 1800°F. Determine the flow rate of methane as the supplementary fuel, if required.

Solution:

a. Assuming that the heat recovery is 70% and the temperature of the waste air from the venting well is 65°F, the temperature of the waste air after the heat exchanger, T_{he}, can be found from Eq. VII.2.9 as

$$T_{he} = \left(\frac{HR}{100}\right)T_c + \left[1 - \frac{HR}{100}\right]T_w = \left(\frac{70}{100}\right)(1800) + \left[1 - \frac{70}{100}\right](65) = 1280°\text{F}$$

b. The average specific heat can be read from Figure VII.2.A as 0.266 Btu/lb-°F at 1800°F.
c. The heat content of the waste gas is 55.6 Btu/lb, as determined in Example VII.2.2.
d. The flow rate of the supplementary fuel can be estimated by using Eq. VII.2.8 as

$$Q_{sf} = \frac{D_w Q_w [C_p(1.1T_c - T_{he} - 0.1T_r) - H_w]}{D_{sf}[H_{sf} - 1.1C_p(T_c - T_r)]}$$

$$= \frac{(0.0739)(200)\{(0.266)[1.1(1800) - 1280 - 0.1(77) - 55.6)\}}{(0.0408)[21,600 - (1.1)[(0.266)(1800 - 77)]} = 2.21\,\text{scfm}$$

VII.2.6 Volume of combustion chamber

The total influent flow to an incinerator is the sum of the waste air, dilution air (and/or the auxiliary air), and the supplementary fuel, and it can be determined by the following equation:

$$Q_{inf} = Q_w + Q_d + Q_{sf} \qquad \text{[Eq. VII.2.10]}$$

where Q_{inf} = the total influent flow rate, scfm.

In most cases, one can assume that the flow rate of the combined gas stream, Q_{inf}, entering the combustion chamber is approximately equal to the flue gas leaving the combustion chamber at standard conditions, Q_{fg}. The volume change across the incineration chamber, due to combustion of VOC and supplementary fuel, is assumed to be small. This is especially true for dilute VOC streams from soil or groundwater remediation.

The flue gas flow rate of actual conditions can be determined from Eq. VII.2.1 or from the following equation:

$$Q_{fg,a} = Q_{fg}\left[\frac{T_c + 460}{77 + 460}\right] = Q_{fg}\left[\frac{T_c + 460}{537}\right] \quad \text{[Eq. VII.2.11]}$$

where $Q_{fg,a}$ is the actual flue gas flow rate in acfm.

The volume of the combustion chamber, V_c, is determined from the residence time, τ (in sec), and $Q_{fg,a}$ by using the following equation:

$$V_c = \left[\left(\frac{Q_{fg,a}}{60}\right)\tau\right] \times 1.05 \quad \text{[Eq. VII.2.12]}$$

The equation is nothing but "residence time = volume ÷ flow rate." The factor of 1.05 is a safety factor, which is an industrial practice to account for minor fluctuations in the flow rate. Table VII.2.B lists the typical thermal incinerator system design values.

Table VII.2.B Typical Thermal Incinerator System Design Values

	Non-halogenated compounds		Halogenated compounds	
Required destruction efficiency (%)	Combustion temperature (°F)	Residence time (sec)	Combustion temperature (°F)	Residence time (sec)
98	1600	0.75	1800	1.0
99	1800	0.75	2000	1.0

From U.S. EPA, Control Technologies for Hazardous Air Pollutants, EPA/625/6-91/014, U.S. EPA, Washington, DC, 1991.

Example VII.2.6 Determine the size of the thermal incinerator

Referring to the remediation project described in Example VII.2.5, an off-gas stream (Q = 200 scfm) containing 800 ppmV of xylene is to be treated by a thermal oxidizer with a recuperative heat exchanger. The combustion temperature is set at 1800°F to achieve a destruction efficiency of 99% or higher. Determine the size of the thermal incinerator.

Solution:

a. Use Eq. VII.2.10 to determine the flue gas flow rate at standard conditions:

$$Q_{fg} \sim Q_{inf} = Q_w + Q_d + Q_{sf} = 200 + 0 + 2.21 = 202.2 \text{ scfm}$$

b. Use Eq. VII.2.11 to determine the flue gas flow rate at actual conditions

$$Q_{fg,a} = Q_{fg}\left[\frac{T_c + 460}{537}\right] = (202.2)\left[\frac{1800 + 460}{537}\right] = 851\,acfm$$

c. From Table VII.2.B, the required residence time is one second. Use Eq. VII.2.12 to determine the size of the combustion chamber as

$$V_c = \left[\left(\frac{Q_{fg,a}}{60}\right)\tau\right] \times 1.05 = \left[\left(\frac{202.2}{60}\right)(1)\right] \times 1.05 = 3.5\,\text{ft}^3$$

VII.3 Catalytic incineration

Catalytic incineration, also known as catalytic oxidation, is another commonly applied combustion technology for treating VOC-laden air. With presence of a precious or base metal catalyst, the combustion temperature is normally between 600 and 1200°F, much lower than that of a direct thermal incineration system.

For catalytic oxidation, the "three T's" (temperature, residence time, and turbulence) are still the important design parameters. In addition, the type of catalyst has a significant effect on the system performance and cost.

VII.3.1 Dilution air

The concentration of flammable vapors to a catalytic incinerator is generally limited to 10 Btu/scf or 135 Btu/lb (equivalent to 20% LEL for most VOCs), which is lower than that for direct incineration. It is due to the fact that higher VOC concentrations may generate too much heat upon combustion to deactivate the catalyst. Therefore, dilution air must be used to lower the contaminant concentration to below 20% of its LEL.

When dilution is required, the volumetric flow rate of the dilution air can be found as

$$Q_{dilution} = \left[\frac{H_w}{H_i} - 1\right]Q_w \qquad \text{[Eq. VII.3.1]}$$

where $Q_{dilution}$ = required dilution air, scfm, Q_w = waste air stream to be treated, scfm, H_w = heat content of the waste air stream, Btu/scf (or Btu/lb), and H_i = heat content of the desired influent entering the treatment system, Btu/scf (or Btu/lb).

Example VII.3.1 Determine the dilution air requirement

Referring to the remediation project described in Example VII.2.3, an off-gas stream (Q = 200 scfm) containing 800 ppmV of xylene is to be treated by a catalytic incinerator with a recuperative heat exchanger. Determine the required dilution air flow rate, if needed.

Solution:

The heating value of the off-gas has been determined as 11.6 Btu/scf or 160 Btu/lb in Example VII.2.3A, which exceeds the 10 Btu/scf or 135 Btu/lb limit. Thus, air dilution is required, and the dilution air flow rate can be determined by using Eq. VII.3.1 as

$$Q_{dilution} = \left[\frac{H_w}{H_i} - 1\right]Q_w = \left[\frac{160}{135} - 1\right](200) = 37\,\text{scfm}$$

Discussion. For the same off-gas, 800 ppmV of xylene, air dilution is required for catalytic incineration but not required for direct incineration.

VII.3.2 *Supplementary heat requirements*

For catalytic incineration of off-gases from soil/groundwater remediation, supplementary heat is often provided by electrical heaters. If natural gas is used, one can use Eq. VII.2.8 to determine the supplementary fuel flow rate. The following two equations should be applied first to estimate the temperature of the flue gas, T_{out}, which would achieve the desired destruction efficiency without damaging the catalyst. It can be estimated from the temperature of the waste gas leaving the heat exchanger (and before entering the catalyst bed), T_{in}, and the heat content of the gas:

$$T_{out} = T_{in} + 50H_w \qquad \text{[Eq. VII.3.2]}$$

On the other hand, the equation can be used to determine the required influent temperature to achieve a desired temperature in the catalyst bed:

$$T_{in} = T_{out} - 50H_w \qquad \text{[Eq. VII.3.3]}$$

where H_w is the heat content of the waste air stream in Btu/scf. These two equations assume a 50°F temperature increase for every 1 Btu/scf of heat content in the influent air to the catalyst bed.

Example VII.3.2 Estimate the temperature of the catalyst bed

Referring to the remediation project described in Example VII.3.1, an off-gas stream (Q = 200 scfm) containing 800 ppmV of xylene is to be treated by a catalytic incinerator with a recuperative heat exchanger. After the heat exchanger, the temperature of the diluted waste gas is 550°F. Estimate the temperature of the catalyst bed.

Solution:

After air dilution, heat content of the diluted waste gas is 10 Btu/scf. Use Eq. VII.3.2 to estimate the temperature of the catalyst bed:

$$T_{out} = T_{in} + 50H_w = 550 + (50)(10) = 1050°\text{F}$$

Discussion. The calculated temperature, 1050°F, falls in the typical temperature range for catalyst beds, i.e., 900 to 1200°F.

VII.3.3 *Volume of the catalyst bed*

The total influent flow to a catalyst bed is the sum of the waste air, dilution air (and/or the auxiliary air), and the supplementary fuel, and it can be determined by the following equation:

$$Q_{\text{inf}} = Q_w + Q_d + Q_{sf} \qquad \text{[Eq. VII.3.4]}$$

where Q_{inf} = the total influent flow rate, scfm.

In most of the cases, one can assume that the flow rate of the combined gas stream, Q_{inf}, entering the catalyst is approximately equal to the flue gas leaving the catalyst at standard conditions, Q_{fg}. The flue gas flow rate of actual conditions can be determined from Eq. VII.2.1 or from the equation below:

$$Q_{fg,a} = Q_{fg}\left[\frac{T_c + 460}{77 + 460}\right] = Q_{fg}\left[\frac{T_c + 460}{537}\right] \qquad \text{[Eq. VII.3.5]}$$

where $Q_{fg,a}$ is the actual flue gas flow rate in acfm.

Because of the short residence time in the catalyst bed, space velocity is commonly used to relate the volumetric air flow rate and the volume of the catalyst bed. The space velocity is defined as the volumetric flow rate of the VOC-laden air entering the catalyst bed divided by the volume of the catalyst bed. It is the inverse of residence time. Table VII.3.A provides the typical design parameters for catalytic incinerators. It should be noted here that the flow rate used in the space velocity calculation is based on the influent gas flow rate at standard conditions, not that of the catalyst bed or the bed effluent.

The size of the catalyst can be determined by

Table VII.3.A Typical Design Parameters for Catalytic Incineration

Desired destruction efficiency (%)	Temperature at catalyst bed inlet (°F)	Temperature at catalyst bed outlet (°F)	Space Velocity (hr^{-1})	
			Base metal	Precious metal
95	600	1000–1200	10,000–15,000	30,000–40,000

From U.S. EPA, Control Technologies for Hazardous Air Pollutants, EPA/625/6-91/014, U.S. EPA, Washington, DC, 1991.

$$V_{cat} = \frac{60Q_{\inf}}{SV} \qquad \text{[Eq. VII.3.6]}$$

where V_{cat} = volume of the catalyst bed, ft^3, Q_{inf} = the total influent flow rate to the catalyst bed, in scfm, and SV = space velocity, hr^{-1}.

Example VII.3.3 *Determine the size of the catalyst bed*

Referring to the remediation project described in Example VII.3.1, an off-gas stream (Q = 200 scfm) containing 800 ppmV of xylene is to be treated by a catalytic incinerator with a recuperative heat exchanger. The design space velocity is 12,000 hr^{-1}. Determine the size of the catalyst bed.

Solution:

a. Use Eq. VII.3.4 to determine the flue gas flow rate at standard conditions:

$$Q_{fg} \sim Q_{inf} = Q_w + Q_d + Q_{sf} = 200 + 37 + 0 = 237 \text{ scfm}$$

b. With a space velocity of 12,000 hr^{-1}, use Eq. VII.2.12 to determine the size of the catalyst bed:

$$V_{cat} = \frac{60Q_{\inf}}{SV} = \frac{(60)(237)}{12,000} = 1.2\,\text{ft}^3$$

Discussion. The size of the catalyst, 1.2 ft^3, is smaller than the volume of the combustion chamber for direct incineration, 3.5 ft^3.

VII.4 Internal combustion engines

The internal combustion (IC) engine of a conventional automobile or truck can be modified and incorporated in a control system to treat VOC-laden air. The IC engine is used as a thermal incinerator, and the physical difference between the IC engine units and the thermal incinerators is mainly in the geometry of the combustion chamber.

VII.4.1 Sizing criteria/application rates

The sizing of an IC engine device is based on the volumetric flow rate of the VOC-laden air to be treated. One vendor reports that their IC engine unit can handle up to 80 cfm of VOC-laden air, while the other reports that their unit can accommodate 100 to 200 scfm of influent gas (depending on the VOC concentrations) for every 300 in^3 of engine capacity.[2] Conservatively speaking, a typical IC engine should not handle more than 100 cfm of VOC-laden air. For a higher flow rate, a treatment system with a few IC engines in parallel would be needed.

Example VII.4.1 Determine the number of IC engines needed

Referring to the remediation project described in Example VII.3.1, an off-gas stream (Q = 200 scfm) containing 800 ppmV of xylene is to be treated by IC engines. Determine the number of IC engines needed for this project.

Solution:

The average off-gas flow rate is 200 scfm, and a typical IC engine can only handle 100 scfm as the maximum. Therefore, a minimum of two IC engines in parallel should be used in this project.

VII.5 Soil beds/biofilters

Biofiltration is an emerging technology for treating VOC-laden air. In biofiltration, the VOC-laden air is vented through a biologically active soil medium where VOCs are biodegraded. The temperature and moisture of the air stream and biofilter bed are critical in design considerations.

VII.5.1 Design criteria

Biofiltration is cost effective for large volume air streams with relatively low concentrations (<1000 ppmV as methane). Maximum influent VOC concentrations have been found to be 3000 to 5000 mg/m^3. For optimum efficiency, the waste air stream should at 20 to 40°C and 95% relative humidity. The filter material should be maintained at 40 to 60% moisture by weight and a pH between 7 and 8. Typical biofilter systems have been designed to treat 1000 to 150,000 m^3/hr waste air with a cross-sectional area of 10 to 2000 m^2. The typical depth of biofilter media is three to four feet.[2] The typical surface loading rate is 100 m^3/hr of waste air stream per m^2 filter cross-sectional area. The required cross-sectional area of the biofilter (A_{filter}) can be determined as

$$A_{biofilter} = \frac{\text{air flow rate}}{\text{surface loading rate}} \qquad \text{[Eq. VII.5.1]}$$

Example VII.5.1 *Size the biofilters for off-gas treatment*

Referring to the remediation project described in Example VII.3.1, an off-gas stream (Q = 200 scfm) containing 800 ppmV of xylene is to be treated by biofilters. Determine the size of the biofilters needed for this project.

Solution:

a. The off-gas contains 800 ppmV of xylene, which is equivalent to 6400 ppmV as methane (each xylene molecule contains eight carbon atoms). This is beyond the typical range of <1000 ppmV as methane. The maximum influent VOC concentrations of 3000 to 4000 mg/m³ have been reported in the literature. Although the xylene concentration in this case (800 ppmV of xylene = 3460 mg/m³) falls within the range, dilution of this off-gas would be a conservative approach. The optimal influent concentration should be determined from a pilot study. In this example, let us dilute the off-gas four times; therefore, the influent flow rate to the biofilter becomes 800 scfm.

b. The typical surface loading rate is 100 m³/hr of waste air stream per m² filter cross-sectional area. Let us convert 800 cfm to m³/hr as

$$Q = 800\ \text{ft}^3/\text{min} = (800\ \text{ft}^3/\text{min})(60\ \text{min/hr})(0.0283\ \text{m}^3/\text{ft}^3)$$
$$= 1360\ \text{m}^3/\text{hr}$$

Use Eq. VII.5.1 to determine the required cross-sectional area as

$$A_{biofilter} = \frac{\text{air flow rate}}{\text{surface loading rate}} = \frac{1360\,\text{m}^3/\text{hr}}{100\,\text{m}^3/\text{hr}/\text{m}^2}$$
$$= 13.6\,\text{m}^2 = 146\,\text{ft}^2$$

c. Use a typical value of 4 feet as the depth of the biofilter.

Discussion. If the biofilter is constructed in a cylindrical shape, the diameter of the biofilter would be around 14 ft.

References

1. U.S. EPA, Control Technologies for Hazardous Air Pollutants, EPA/625/6-91/014, U.S. EPA, Washington, DC, 1991.
2. U.S. EPA, Control of Air Emissions for Superfund Sites, EPA/625/R-92/012, U.S. EPA, Washington, DC, 1992.

Index

Abiotic degradation 88
Absorption 36
Acetone 83, 201, 231
Activated carbon 198-205, 229-237
  liquid phase 230
  vapor phase 230
Activated carbon adsorber 3, 198-205, 229-237
  adsorption capacity 199-201, 230-233
  adsorption isotherm 199-201, 230-233
  configuration 202
  contaminant removal rate 202, 234-236
  cross-sectional area 201, 233
  empty bed contact time (EBCT) 201
  height 202, 233
  regeneration 202, 236-237
Adhesion 174
Adsorbate 42
Adsorbent 42, 198
Adsorption efficiency 229
Adsorption isotherm 42-43, 199-201, 230-233
Adsorption 42, 88, 140
Advanced oxidation process (AOP) 211-213
Advection–dispersion equation 81-82, 95
Advective flow 99
Advective force 57
Aerobic conditions 168, 169
Air emission control 229-253
Air flow rate 164-165
  actual 238
  standard 238
Air infiltration 139
Air sparging 222-227
  air injection pressure 223
  oxygen transfer efficiency 222
Air stripping 205-211
  air-to-water ratio 206
  column diameter 208
  diffused aeration 205
  flooding 207
  minimum air-to-water ratio 206
  packed column (towers) 205
  packing height 206, 208
  spray systems 205
  stripping factor 206
  tray columns 205
Alcohol 175
Amines 84
Anaerobic condition 168
Anthracite 84
Antoine constant 33
Antoine equation 33
Aquifer
  artesian 67
  confined 65, 184
  unconfined 20, 22, 65, 187
  water table 71
Aquifer heterogeneity 82

Aquifer test 6, 68, 74-81
Cooper–Jacob straight-line method 77-79
Distance-drawdown method 79-81
Theis curve matching 75-77
Aquifer thickness 64, 68
Aromatics 45
Attached-growth system 214-215
Auto-ignition temperature 238
Auxiliary fuel 238, 240, 243 244

Batch reactors 105, 112, 113-117
design equations 113
Bedrock 6
Bench-scale testing 115
Bentonite seal 29-31
Benzene 8, 10, 11, 18, 34, 35, 39, 83, 231
Benzoic acid 83
Biodegradation 139
Biofiltration 3
Biological process 88, 103
Biological reactors
attached growth 214-215
suspended growth 214
Biological treatment 214-221
Biomarker 18
Bioremediation 1
groundwater 214-221
soil 168-174
oxygen supply 217-220
nutrient addition 220-221
Biotic degradation 88
Blower, *see* vacuum blowers
Boiling point 83
Breakthrough, adsorption 199
Bromine 84
Bromomethane 39
BTEX 18
BTU value, *see* heating value
Bulk density 8, 9, 12, 15, 88, 144, 146, 199
Butanol 83
2-Butanone 39

Capillary fringe 5, 20-22, 222
Capillary pressure 222
Capillary rise 20, 21
Capillary zone, *see* capillary fringe
Capture zone analysis 183, 189-198
downstream distance 191, 195
side-stream distance 191, 195
Carbon adsorption, *see* activated carbon
Carbon atom 84
Carbon disulfide 39
Carboxylic acids 45
Catalysts
base metal 251
precious metal 251
Catalytic incineration 3, 248-251
deactivation of catalyst 248
dilution air 248
space velocity 251
supplementary heat 249
volume of catalyst bed 250
Cation exchange capacity
Caustic soda 213
Chemical kinetics 107-112
rate equations 107- 109
rate of conversion 107
reaction order 107
reaction rate constant 107
Chemical process 88, 103
Chlorinated hydrocarbons 45
Chlorine 84
Chlorobenzene 39, 231
Chloroethane 39
Chloroform 39
Chloromethane 39
Clay 61
hydraulic conductivity 64
intrinsic permeability 64
porosity 65
specific retention 65
specific yield 65
Clayey aquifer 64, 79
Clausius–Clapeyron equation 33
Clay liner 62
Cleanup time 158-164
Coefficient of permeability, *see* permeability
Combustion temperature 237, 245, 247, 248
Compaction 174
Completely mixed flow (CMF) reactor, *see* continuous flow stirred tank reactor
Completely stirred tank reactor (CSTR), *see* continuous flow stirred tank reactor

Compressors, *see also* vacuum blowers 166
  reciprocating 167
Concentration gradient 57, 82
Conductivity 31
Confined aquifer, *see also* aquifer
  steady-state flow 67-71, 184-187
Cone of depression 193, 184-189
Confined space 36
Conservation of mass 104
Contact face 59
Contaminant concentration 8
  relationship with mass 811
Contaminant properties 39-40
Contaminant removal rate 155-158
Contaminated groundwater 16
Contaminated soil
  excavated 2
  left in the vadose zone 2
  mass 15-17
  volume 15-17
Contaminated plume, size of 2
Continuous flow stirred tank reactors (CFSTR) 112, 117-119, 179
  design equations 118
Conversion factors 8
Cooper–Jacob straight-line method 77-79
Corrective action 5
Cyclohexane 231

Darcy unit 62
Darcy's law 58-59, 94
Darcy's velocity 59-61, 151
Decay rate 111
Detection limits 14
Degree of water saturation 144, 146, 169
Dense non-aqueous phase liquids (DNAPLs), *see also* non-aqueous phase liquids 6
Density 62
  of water 62-63
  of air 240
Design consideration
  activated carbon adsorption 202
  air stripping 205-211
  catalytic incineration 248-251
Desorption 174, 175, 178
Destruction and removal efficiency (DRE) 238
Detection limit 14
Detention time, *see* residence time
Dibromochloromethane 39
Dibromomethane 39
1,2-Dichlorobenzene 45
1,1-Dichloroethane 39, 77
1,2-Dichloroethane 39, 77
1,1-Dichloroethylene 39
1,2-Dichloroethylene 39
1,2-Dichloropropane 39
1,3-Dichloropropylene 39
Diesel fuel 18
Diffusion, molecular 57, 82-88
Diffusion coefficient 82-88
  air 96
  values in air 39-40
  values in water 83
Digestive system 10, 11
Direct incineration 3
Discharge velocity 59
Dispersion
  coefficient 82-88, 89, 95
  hydraulic 82
Dispersivity 83
Dissolution 36, 88
Dissolved groundwater plume 49, 57
  migration 90-94
Distance-drawdown method 79-81
Downgradient well 90
Drawdown 6, 68, 69, 74, 75, 77, 79, 148
Dulong's formula 240
Dwell time, *see* residence time

Elutriate 175
Empty bed contact time (EBCT) 201
Enthalpy of vaporization 33
EPA method 418.1 18
EPA method 8015 14, 18
EPA method 8020 18
Equilibrium
  free product and vapor 33-36
  liquid and vapor 36-42
  solid and liquid 42-46
  solid, liquid and vapor 46-48
Equalization tank
Equipotential lines 66, 67
Esters 45, 84
Ethers 84
Ethylbenzene 18, 39
Ethylene glycol 83

Explosive limit
- lower 241-242
- upper 241-242

Extent of plume 57
- areal 6, 14
- vertical 6, 14
- determination 7-28

Exothermic reaction 230
Extraction well 148

Factor of safety 30, 148
Fate and transport 6, 43, 57
Feasibility study 170
Fence diagram, contaminant 15
Fick's law 95-96
Financially responsible parties 1
First order reactions 108, 113, 118, 120, 211
Flexible membrane liners 63
Flooding, air stripping 207
Flow lines 82
Flue gas 246, 247
Fraction of organic carbon 44, 144
Free floating product, *see* free product
Free product 5, 18, 32, 48, 57, 140, 155
- mass 22-24
- volume 22-24
- actual thickness 22-24
- apparent thickness 22-24

Freundlich isotherm 43, 199
- empirical constants 231

Fully penetrating well 67-68, 71, 184, 187

Gas chromatograph 18
Gasoline 17
- fresh 141
- molecular weight 19
- new-formula 18
- weathered 141

Gas-phase transport 95
Grain size analysis 74
Granular activated carbon (GAC), *see* activated carbon
Gravel
- hydraulic conductivity 64
- intrinsic permeability 64
- porosity 65
- specific retention 65
- specific yield 65

Gravel pack, monitoring well 29-31
Gravitational constant 62
Gravity potential 94
Groundwater elevation 6, 66
Groundwater extraction, *see* hydraulic control
Groundwater flow
- gradient 6, 66-67, 92
- direction 6, 66-67
- velocity 59, 89

Groundwater monitoring wells 6, 28-32
- casing 31
- well volume 31-32
- well water depth 31

Groundwater movement 2, 58-67
Groundwater plume 217
- age 2

Groundwater pumping 67-74
- confined aquifer 67-71
- unconfined aquifer 71-74

Groundwater remediation 66, 183-228
- capture zone analysis 189-198
- capture zone 2, 183
- cone of depression 183
- ex situ 2, 198-216
- extent of contamination 5
- hydraulic control 183-198
- in situ 2, 217-227
- optimal well spacing 2, 197-198

Groundwater remediation techniques
- activated carbon adsorption 2, 198-105
- advanced oxidation process 2, 211-213
- air sparging 2, 222-227
- air stripping 2, 205-211
- bioremediation 2, 214-217
- chemical precipitation 213-214

Groundwater sampling
- purge volume 31-32
- purging 6
- well volume 31-32

Grout 30

Half-life 110-112
Halogenated compounds 45, 247
Head loss 58
Health risk 18, 57
Heat capacity 226, 245
Heat content, *see* heating value

Heat exchanger 245, 249
Heat recovery 245
Heating value 238, 239-241, 249
Heavy metals 81, 175, 213
Height of transfer unit (HTU) 209
Henry's constant 36-42, 44, 100, 144, 146, 206
    conversion table 37
    values 40-41
Henry's law 36-42, 206
Horsepower requirements 166
    actual 167
    theoretical 167
Hydraulic conductivity 6, 58, 61-64, 68-81, 94
    at saturation 94
    conversion factors 59
    relative 94
    typical values 64
Hydraulic control 183-198
Hydraulic gradient 58, 59, 61, 64
Hydraulic loading rate 205
Hydraulic retention time, *see* residence time
Hydrocarbons
    heavier end 18
    lighter end 18
Hydrogen peroxide 211, 217
Hydrophilic compounds 44
Hydrophobic compounds 44, 90, 100, 201
Hydrostatic pressure 223

Ideal gas 8, 166
Ideal gas law 238
Ideal reactors 113
Ideal solution 34
In situ vacuum extraction,
    see also soil venting 139
In situ volatilization,
    see also soil venting 139
Indigenous microorganisms 217
Infiltrating water 57
Infiltration gallery 217
Infrared (IR) 18
Ingestion system 10, 11
Inhalation system 10, 11
Insecticides 45
Internal combustion (IC) engines 251-252
Interstitial velocity 60
Ionization 88

Isentropic compression 166
Isothermal compression 166

Kinematic viscosity 61
Knockout drums 140

Laminar flow 58
Landfill 58, 63
    hazardous waste 11
    sanitary 11
Langmuir isotherm 43, 199
LeBas method 84
Linear adsorption isotherm 43, 88
Liquefied butane 175
Liquefied propane 175
Longitudinal mixing 119
Lower explosive limit (LEL),
    see explosive limit
Low-temperature heating 178-181

Mass balance 43
    concept 2, 81, 104-107
    equation 104-106, 117, 119, 131, 175, 206
Mass fraction 17-20
Mass loading rate 105, 235
Mass of contaminants, determination of
    8-11
    adsorbed to soil 48
    in air 9
    in liquid 8
    in pore void 48
    in soil 8
    in soil moisture 48
Mass transfer 156
Material balance, *see* mass balance
Meinzer unit 63
Metabolic product 139, 168
Metal oxides and hydroxides 61
Metering pump 211
Methylene chloride 10, 39
Microorganisms 168
    environmental conditions for
    growth 169
Migration speed, dissolved plume 81-94
Millington–Quirk equation 96
Moisture content 169

Moisture requirement 168-170
Moisture potential 94
Moisture removal device,
see knock-out drum
Mole fraction 17-20
Molarity 37
Molar volume 8, 83, 84
additive-volume increment 84
Molecular diffusion, *see also* diffusion 82
coefficient 82
effective 82
Molecular weight 84
gasoline 141
Monterey sand 29

Naphthalene 84
Natural gas 84
Nonaqueous phase liquids (NAPLs) 5, 32, 46
Number of transfer unit 209
Nutrient requirements 169, 170-172

Observation well 75, 79, 148
Octanol water partition coefficient 44-45, 144, 146
values 39-40
Off-gas 140, 178
Off-gas treatment,
see VOC-laden air treatment
Optimum nutrient ratio 170
Organic carbon partition coefficient 44-45
Organic content of soil,
see fraction of organic carbon
Organic solvent 175, 177
Oxidation–reduction potential 170
Oxygen demand 222
Oxygen 217, 222
Oxygen requirements 172-174
Ozone 211

PAHs 45
Partial pressure 33-42, 140
Partition coefficient 43-45, 89, 175
Partitioning 32-55, 139
contaminant mass 2, 6
Part per million (ppm) 7
ppm by volume (ppmV) 7

Passive venting wells 139
PCBs 45, 201
PCE, *see also* tetrachloroethylene 45
Perchloroethylene (PCE)
see tetrachloroethylene
Permissible exposure limit (PEL) 10
Permeability, intrinsic 61-63
typical values 64
Permeability, relative 94
Pesticides 45
Phenol 231
Physical treatment 103
Phosphorus 170
Pilot studies 115, 149, 178
Plug flow 112, 211
Plug flow reactors 112-113, 119-122
Plume
migration 5, 183
location 14
extent 14
dissolved 20
Poise 62
Polynuclear aromatic hydrocarbons,
see PAHs
Pore radius 20
Pore geometrics 82
Porosity 16, 60, 64
air-filled 96
total 96
Potential responsible parties 18
Precipitation, chemical 213-214
Preliminary design 103, 140
Pressure head 94
Pressure profile 148-151
Process flow diagram 104
Product age 18
Propanol 83
Pump and treat 94, 183
Pumping rate 68, 74, 75
Pumping test,
see also aquifer test 74 -75
suggested guidelines 81
Purge volume 31
Pyrene 39, 53

Radius of influence 148-151
Raoult's law 34, 140
Reaction
kinetics 2, 108
order 2, 108

rate constant 2, 108, 179

Reactors
- batchwise 112-117
- CFSTR 112-119
- continuous flow 112
- PFR 113, 113, 119-122
- sizing 2, 122-123
- steady state 112
- types 2, 112-119
- unsteady state 112

Reactor configuration 124-137
- in parallel 131-137
- in series 125-131

Reactor design 2, 103-137
Reactor volume 108
Regeneration, carbon 230
Relative humidity 230
Remedial investigation 5
Remedial alternatives 5, 15
Removal efficiency 2
Removal mechanisms 88, 109, 174
Residence time 2, 113, 122
Retardation factor 88-94
- vapor migration 98-100

Retention time, *see* residence time
Risk assessment 18
Run-on control 168
Run-off control 168

Safety factors, *see* factor of safety
Sample collection
- groundwater 6
- soil 6

Sand
- hydraulic conductivity 64
- intrinsic permeability 64
- porosity 65
- specific retention 65
- specific yield 65

Saturated dissolved oxygen concentration 172
Saturated zone 95, 222
Second order reaction 108, 113, 118, 120
Seepage velocity 59-61
Semi-volatile compounds 178
Silt
- hydraulic conductivity 64
- intrinsic permeability 64

Silty sand
- hydraulic conductivity 64
- intrinsic permeability 64

Site assessment 1, 2
Site characterization,
- *see also* site assessment 5

Site remediation 1, 31
Site restoration 1
SI units 3
Slug test 74-75
Slurry reactors, soil 118, 168
Soil
- porosity 65
- specific retention 65
- specific yield 65

Soil biofilters 252-253
Soil bioremediation 168-174
- moisture requirement 168- 170
- nutrient requirement 170-171
- oxygen requirement 172-174

Soil boring 6, 15, 28-31
- cuttings 6, 28-31

Soil characteristics 6
Soil excavation 11, 13
Soil expansion 13
Soil fluffy factor 12, 13, 28
Soil flushing 174
Soil mechanics 63
Soil moisture 32, 42, 95, 99, 139
Soil organic matter 61
Soil remediation, vadose zone 139-181
- aboveground 2
- extent of contamination 5
- in situ 2
- low temperature heating 181
- soil bioremediation 168-174
- soil flushing 175
- soil vapor extraction 139- 168, 222
- soil washing 174-178
- solvent extraction 175

Soil slurry reactor 118
Soil stockpiles 11-14
Soil vapor extraction,
- *see also* soil venting 139

Soil venting 1, 139-168
- cleanup time 2, 158-164
- effect of temperature 2, 164-165
- extracted contaminant concentration 2, 140-147
- pressure profile 148-151

radius of influence 148-151
radius of influence 2
sizing of blowers 2, 166-168
vapor flow rate 2, 151-155
well spacing 2, 165
Soil vapor extraction, *see* soil venting 139
Soil vapor stripping, *see* soil venting 139
Soil washing 2
Soil water pressure head 94
Solidification/stabilization 229
Solubility 37, 44-45
values 39-40
Solute 83
Solvent extraction 174
Solvents 42
Space velocity 251
Specific capacity 68, 69, 71
Specific gravity 10
Specific gravity 37
Specific retention 64-65
Specific yield 64-66
Stagnation point 211
Standard conditions 238, 246, 250
Standard flow rate 152
Static head 68
Static mixer, in-line 211
Static soil pile 168
Steady-state conditions 105, 106, 112, 117, 119
Stoichiometric relationship 172
Storage coefficient,
see also storativity 64, 73
Storativity 64-66, 74, 75, 77, 79
Stripping factor,
see also air stripping 206-207
Stripping, *see* air stripping
Styrene 39
Surface loading rate 215
Supplementary fuel 238, 245-246

Tank pit 11-14
Tank removal 6, 11-14
Temperature, effect on bioremediation 169
1,1,1,2-Tetrachloroethane 40
1,1,2,2-Tetrachloroethane 40
Tetrachloroethylene 40
Tetrachloromethane 40
Theis curve matching 75-76
Thermal incinerator,
see thermal oxidizer
Thermal oxidation 237-246
air flow rate 238
auxiliary air 243
dilution air 241
heating values 239
supplementary fuel 245
temperature 238
volume of combustion chamber 246
Thermal oxidizer 238, 241
air flow rate vs. temperature 238-239
auxiliary air 243-244
dilution air 241-243
heating value 239-241
supplementary fuel 245-246
typical design values 247
volume of combustion chamber 246-248
Thermal process 103, 237-252
Three "T's" of combustion 237-238, 248
Toluene 18, 40, 231
Tortuosity factor 83
air-phase 96
Total petroleum hydrocarbon (TPH) 13, 14, 18
Transmissivity 64-66, 69, 75, 77. 79
Treatment process, selection of 103
Tribromoethane 40
1,1,1-Trichloroethane 40
1,1,2-Trichloroethane 40
Trichloroethylene 40
Trichlorofluoromethane 40
Trickling filters 214-215
Turbulence 238

Ultraviolet irradiation 211
Unconfined aquifer, *see also* aquifers
steady-state flow 71-74, 187-189
Underground storage tanks (USTs) 5, 11-14
Unit operation/process 103
Universal gas constant 33
Unsaturated zone, *see* vadose zone
Upper explosive limit (UEL),
see explosive limit
U.S. customary units 3

Vacuum extraction, *see* soil venting
Vacuum pump (blower) 139
  sizing 166-168
Vadose zone 5, 32, 46, 57, 222
  contaminant transport 94-100
Vadose zone soil remediation 139-182
  low temperature heating 178-181
  soil bioremediation 168-174
  soil vapor extraction 139-168
  soil washing 174 -178
Vapor concentration 32, 140-148, 155
Vapor pressure 37, 140
  values 39-40
  gasoline 141
Vinyl chloride 40, 231
Viscosity 62, 83, 84
  of water 63
VOCs 57, 81, 139
VOC-laden air treatment
  activated carbon adsorption 3, 229-237
  biofiltration 3, 252-253
  catalytic incineration 3, 248-251
  direct incineration 3, 237-246
  IC engines 3, 251-252
Volatile organic compounds, *see* VOCs
Volatilization 33, 88, 139, 140, 178, 205
Volumetric flow rate 58
Volumetric water content 169

Water
  density values 63
  viscosity values 63
Water content, volumetric 94
Water table 20, 21, 22, 67
  elevation 66, 67
Well discharge rate,
  see also pumping rate 68
Well function 76
Well spacing
  soil vapor extraction 165
  groundwater extraction 197-198
Well volume, groundwater sampling 31, 32
Well yield, *see also* pumping rate 68
Wilke–Chang method 83

Xylenes 9, 18, 40, 231

Zeroth order reaction 108, 113, 118, 120